Pythons of Australia

BRIAN A. KEND

Edited by

Gordon W. Schuett, PhD
Arizona State University West
Phoenix, Arizona
and
Laura White Schuett, MA
Glendale Community College
Glendale, Arizona

Canyonlands Publishing Group, LC
Provo, Utah, USA

Original Author Contract by HERP · GOODS

ISBN 0-9657446-1-2

Library of Congress Catalogue Card Number: 97-69689

Printed in the United States of America

Graphic Design and Color Scans:
John L. Crawley &
Michael W. Hancock
Focus Design
580 South State Street
Orem, Utah 84058
(801) 227-0759

To my grandfather, Dr. Milton Rowen,

for his love of wildlife and efforts in conservation.

Foreword

While we may rightly marvel at our technological accomplishments, especially in the field of outer space exploration, an irony persists: we know very little about the secret lives of animals, and in particular, the tropical snakes with which we share the planet. What do they do? When do they do it? Why? These questions have haunted observant students and naturalists throughout remembered time.

The pursuit of natural history as an important component of science was sanctioned and popular—in the 19th century. How unfortunate that today, while priceless habitats worldwide erode, we have chosen to ignore the pressing need to document and understand how animals go about the business of living. But there is hope: the traditional reliance of natural history upon anecdote, anathema to modern science, has recently been recognized as a valuable resource. A new journal, *Herpetological Natural History*, is providing a much-needed outlet for these observations. International interest in conservation of rainforests has focused attention on the plight of tropical animals. And great books like Rick Shine's *Australian Snakes, A Natural History,* and Harry Greene's *Snakes, The Evolution of Mystery in Nature* are beginning to appear.

Our eternal preoccupation with classification notwithstanding, the least interesting aspect of an animal is its name. Knowing that a python is called *Antaresia perthensis* pales when compared with knowing where it hides, when it basks, what it eats and how it forages, when it breeds—in short, how it lives. *Pythons of Australia* represents an attempt, through compilation of anecdotes, published data, and photographs of natural events taken as they occured, to provide the reader with an insight and appreciation for these remarkable snakes as whole animals. It represents the experiences and generous contributions of many people, presented in a straightforward and objective manner.

Through writing this book, Brian Kend, an avid student of pythons, has accomplished more than he intended. His perseverance during what by all accounts was an arduous task has produced a fresh and interesting contribution. But perhaps more importantly, he has done what many fail to do: he has given something back to the field of herpetology.

William W. Lamar
School of Sciences
University of Texas at Tyler
May, 1997

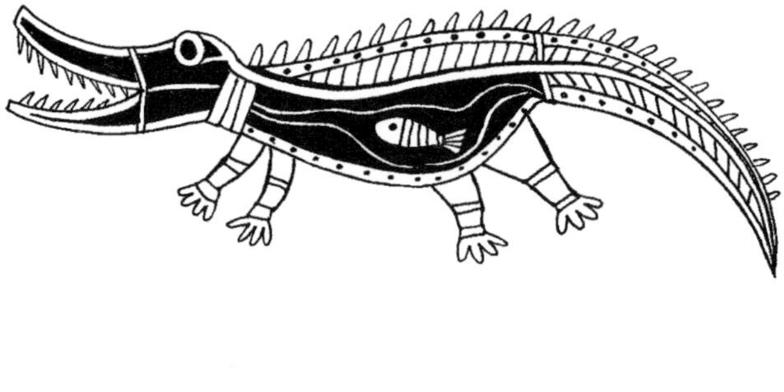

Acknowledgments

This book would not have been possible without the generous contributions of many people who answered my letters requesting information on python distribution and biology. I am particularly thankful to Australian herpetologists and naturalists, especially Gavin Bedford, Rob Bredl, Harold Cogger, Peter Comber, Robert Browne-Cooper, Brian Bush, Neil Charles, Bryan Cantle, Jeanette Covacevich, Greg Czechura, Steve Donnellan, Ray Field, Clifford Frith, Dawn Frith, Greg Fyfe, Graeme Gow, Greg Harold, Raymond Hoser, Steve Irwin, Terri Irwin, Ron Johnstone, Simon Kortlang, Peter Jones, Brad Maryan, Greg Miles, Matthew Milgate, Ian Morris, Lyall Naylor, Magnus Peterson, Richard Shine, Laurie Smith, John Weigel, Bob Withey, and Steve Wilson. Outside of Australia, I am indebted for the invaluable assistance provided by Allen Anderson, David Barker, Tracy Barker, Corey Blanc, Steve Bostwick, Gary Braddock, Dale DeNardo, Philippe de Vosjoli, Harvey Fischer, Norman Frank, Arnold Kluge, William Lamar, Casey Lazik, William Love, Paul Lynum, Jerry Marzec, Scott Michaels, Paul Miles, Matthew Moyle, Freek Nuyt, Bruce Phillips, Louis Porras, Doug Price, Holly Price, Ray Price, Richard Ross, Gordon Schuett, Laura Schuett, Tom Schultz, Terry Schwaner, Gary Sipperley, and Trooper Walsh.

For reviewing the manuscript and their many helpful comments and suggestions I thank Neil Charles, Harold Cogger, Dale DeNardo,

Ray Field, Harvey Fischer, Greg Fyfe, Clay Garrett, Raymond Hoser, Steve Irwin, Terri Irwin, Ron Johnstone, Price Keeler, William Lamar, Brad Maryan, Louis Porras, Richard Ross, Gordon Schuett, Laura Schuett, Richard Shine, Laurie Smith, Bern Tryon, John Whitehead, and Bob Withey.

For assistance with range maps, taxonomic keys, and discussion regarding pythonid systematics, I am grateful to Steve Donnellan, Greg Fyfe, Arnold Kluge, William Lamar, Brad Maryan, Gordon Schuett, Laurie Smith, and Bob Withey. My brother, Scott Kend, provided necessary computer support, especially in generating the range maps and graphs.

Slides, photographs, and illustrations were generously provided by Brian Barnett, Corey Blanc, Steve Bostwick, Rob Bredl, Carel P. Brest van Kempen, Robert Browne-Cooper, Brian Bush, Neil Charles, Bryan Cantle, Peter Comber, Jeanette Covacevich, Greg Czechura, Dale DeNardo, Ray Field, Clifford Frith, Dawn Frith, Greg Fyfe, Greg Harold, Carl Hess, Matt Hingley, Paul Horner, Raymond Hoser, Grant Husband, Steve Irwin, Terri Irwin, Ron Johnstone, Casey Lazik, William Love, Brad Maryan, Greg Miles, Matthew Milgate, Ian Morris, Lyall Naylor, Magnus Peterson, the Queensland Museum, Terry Schwaner, John Tashjian, Brad Tylman, Trooper Walsh, Tom Weidner, Steve Wilson, Bob Withey, and John Young.

I am especially thankful to Gordon Schuett and Laura Schuett for the time spent in editing the manuscript and the many courtesies extended throughout the preparation of this book. To my friend, Louis Porras, I owe a debt of gratitude that I only hope someday to repay. To my devoted parents, Dr. and Mrs. Steven Kend, and my brother Scott Kend, a very special and warm acknowledgment for their patience, understanding, and encouragement when this work at times seemed interminable. Finally, I am indebted to Greg Fyfe and Bob Withey for the wealth of information they provided on python ecology. In most cases their contributions have been included freely in the text. No acknowledgement here can fully express my gratitude.

Chapter 1

Introduction

Evolution, Radiation, and Distribution of Australian Pythons

Australia is home to well-known and popular marsupials such as the Red Kangaroo, Koala, and Common Wombat. Perhaps lesser known is that nearly 1000 species of amphibians and reptiles are found in Australia. The diverse herpetofauna ranges from colorful treefrogs to unusual geckos and dangerously venomous snakes. Included in this wide range of diversity are the pythons.

The evolutionary origin of pythons is a subject that remains under debate. One hypothesis is that pythons or their ancestors arrived in Australia from the north in the last 15-20 million years when tectonic plate movements caused the island to collide with the Asian mainland. According to Underwood & Stimson (1990), the Australia-New Guinea radiation of pythons originated from southeast Asia. These early forms may have resembled the large pythons presently found in Asia (Shine, 1991). Kluge (1993) has proposed, based on his phylogenetic and biogeographical studies of extant pythons, that they had their origin in the Australia-New Guinea region, and that species from southeast Asia and Africa (genus *Python*) evolved from this radiation (Fig. 1).

Fossils of extinct pythons from the Miocene Riversleigh deposits in northwestern Queensland are estimated to be 12-20 million years old. This evidence suggests that pythons were established in Australia prior to that time.[5] Fossils of the humorously named *Montepythonoideis riversleighensis* and *Morelia antiqua* were reported by Kluge (1993) likely to resemble the extant *Morelia spilota* and *Liasis olivaceus*, respectively, and he placed them in synonymy.

The radiation of pythons in Australia is believed to have occurred rather quickly (Smith & Plane, 1985), becoming established throughout New Guinea, Indonesia, and the western Pacific (McDowell 1975, 1979; Kluge, 1993) (Fig. 2). The Papuan Python (*Apodora papuana*), Ringed Python (*Bothrochilus boa*), Indonesian Water Python (*L. m. mackloti*), Savu Island Python (*L. m. savuensis*), and Boelen's Python (*M. boeleni*), as well as the five species shared with Australia, viz., the White-Lipped Python (*L. albertisii*), Water Python (*L. fuscus*), Scrub Python (*M. amethistina*), Green Python (*M. viridis*), and New Guinea Carpet Python (*M. s. variegata*), colonized similar habitats as their Australian relatives.

Pythons have evolved in and adapted to a variety of habitats, but nowhere is this better observed than in Australia, where they inhabit regions ranging from desert sand dunes to sandstone rock escarpments and tropical rainforests. They clearly are an extremely successful and diverse group of snakes.

Python Systematics

In recent years there have been several changes in pythonid nomenclature. Problems are evidenced by continued revisions in classification (Stimson, 1969; McDowell, 1975; Smith, 1981a,b; Cogger et al., 1983; Wells & Wellington, 1984; Smith, 1985; Wells & Wellington, 1985; Underwood & Stimson, 1990; Kluge, 1993; Barker & Barker, 1994b).

I have relied primarily on the revision of Kluge (1993), who recognized 18 species of pythons in Australia assigned to the genera

Antaresia, Aspidites, Leiopython, Liasis, and *Morelia.* His approach was cladistic (= phylogenetic systematics) which focuses on genealogical relationships (Wiley, 1981). All taxa in Kluge's (1993) work were recognized as binomials (i.e., the genus followed by the specific epithet). To ensure clarity and provide the reader with a more balanced perspective, I have recognized the following taxa: *A. stimsoni orientalis, L. fuscus, L. mackloti mackloti, L. mackloti savuensis, L. olivaceus barroni, M. spilota spilota, M. spilota bredli, M. spilota imbricata,* and *M. spilota variegata.*

Because of the lack of a thorough systematic treatise in addressing interpopulational relationships within *M. spilota*, I have followed the proposed combinations of Wells & Wellington (1984, 1985) and Barker & Barker (1994b). This does not imply that I am in complete agreement with this arrangement, but for the purpose of this book, delineating the populations of *M. spilota* appears to make biological sense (see **Interpreting Relationships in the *M. spilota* Group**).

With the inclusion of the subspecies of *Morelia spilota*, i.e., *M. s. cheynei*, *M. s. mcdowelli,* and *M. s. metcalfei*, Australia has 22 taxa of pythons. Differences in the treatment of taxonomy between this book and Kluge (1993) can be noted by comparing the **Index of Common and Scientific Names** with **A Checklist of Australian Pythons**, respectively. It is likely that in the future, however, some of the above populations will be evaluated and reclassified (e.g., elevated to specific status). This is due, in large part, to the ongoing controversy regarding the subspecies concept (Wiley, 1981; Kluge, 1991; 1993).

Prior to Kluge's (1993) work, the most diverse group of pythons was *Liasis*, which included both the *childreni* and *olivacea* groups. He partitioned *Liasis* into two genera: *Antaresia*, which includes the five members of the *A. childreni* group, and *Liasis*, which includes the three members of the *L. olivacea* group. In addition, the semi-fossorial genus *Aspidites*, which has two species, was found to be the basal lineage of all pythons.

Another geographically widespread group is *Morelia*, which is presently comprised of 11 taxa, including the first described Australasian species, *M. amethistina* (Schneider, 1801), two recently discovered species, *M. oenpelliensis* (Gow, 1977) and *M. carinata* (Smith, 1981a), and the various populations of *M. spilota*. The Green Python, formerly in the monotypic genus, *Chondropython*, was placed in the genus *Morelia* (Kluge, 1993). The White-Lipped Python, recently placed in the genus *Morelia* (Underwood & Stimson, 1990), was placed in the monotypic genus, *Leiopython* (Kluge, 1993).

Index of Common and Scientific Names

Genus *Antaresia*

Northern Brown Python	*Antaresia childreni*
Spotted Python	*Antaresia maculosus*
Pygmy Python	*Antaresia perthensis*
Western Stimson's Python	*Antaresia stimsoni stimsoni*
Eastern Stimson's Python	*Antaresia stimsoni orientalis*

Genus *Aspidites*

Black-Headed Python	*Aspidites melanocephalus*
Woma	*Aspidites ramsayi*

Genus *Leiopython*

White-Lipped Python	*Leiopython albertisii*

Genus *Liasis*

Water Python	*Liasis fuscus*
Olive Python	*Liasis olivaceus olivaceus*
Pilbara Olive Python	*Liasis olivaceus barroni*

Genus *Morelia*

Scrub Python	*Morelia amethistina*
Rough-Scaled Python	*Morelia carinata*
Oenpelli Python	*Morelia oenpelliensis*
Diamond Python	*Morelia spilota spilota*

Central Carpet Python	*Morelia spilota bredli*
Jungle Carpet Python	*Morelia spilota cheynei*
Southwestern Carpet Python	*Morelia spilota imbricata*
Eastern Carpet Python	*Morelia spilota mcdowelli*
Inland Rivers Carpet Python	*Morelia spilota metcalfei*
Northern Carpet Python	*Morelia spilota variegata*
Green Python	*Morelia viridis*

A Checklist of Australian Pythons

In assembling a checklist of Australia's pythons, a synonymy of alternatives of both common and scientific names, which have appeared in the literature, is included for each taxon. The synonymy begins with the original citation and ends with the current common and scientific names I accept (**bold**). For many of the common names, I borrowed extensively from Kinghorn (1964), Shine (1991), Cogger (1992), Ehmann (1992), Pearson (1993), and Barker & Barker (1994b). As mentioned earlier, subspecies have not been included due to the difficulty in defining subspecific populations and the changing view of the species concept (Wiley, 1981; Kluge, 1991; 1993).

The purpose of this list is to provide background information for those unfamiliar with pythonid nomenclature, specifically the revisions proposed by Kluge (1993). For more information, also see Stimson (1969); McDowell (1975; 1979); Smith (1981a,b; 1985); Cogger, Cameron, & Cogger (1983); Wells & Wellington (1984; 1985); Stafford (1986); Underwood & Stimson (1990); Kluge (1993); and Barker & Barker (1994b).

Antaresia Wells & Wellington, 1984
Liasis Childreni (Gray, 1842)
Banded Rock Python, Children's python, **Northern Brown Python**
    Nardoa Gilbertii - Gray, 1842
    Liasis childreni - Stimson, 1969

Bothrochilus childreni - Cogger et al., 1983
Antaresia gilbertii- Wells & Wellington, 1984
Morelia childreni - Underwood & Stimson, 1990
Antaresia childreni - Kluge, 1993

Liasis maculosus (W. Peters, 1873)
Banded Rock Python, Children's Python, Eastern Children's Python, Small-Blotched python, **Spotted Python**
Antaresia maculosus - Wells & Wellington, 1984
Liasis maculosus - Smith, 1985
Morelia maculosus - Underwood & Stimson, 1990
Antaresia maculosus - Kluge, 1993

Liasis childreni perthensis (Stull, 1932)
Perth Pygmy Python (erroneous), Western Children's Python, Ant-Hill Python, **Pygmy Python**
Liasis perthensis - Stimson, 1969
Bothrochilus perthensis - Cogger et al., 1983
Antaresia perthensis - Wells & Wellington, 1984
Morelia perthensis - Underwood & Stimson, 1990
Antaresia perthensis - Kluge, 1993

Liasis stimsoni (Smith, 1985)
Banded Rock Python, Large-Blotched Python, **Stimson's Python**
Antaresia childreni - Wells & Wellington, 1984
Antaresia saxacola - Wells & Wellington, 1985
Liasis stimsoni - Smith, 1985
Morelia stimsoni - Underwood & Stimson, 1990
Antaresia stimsoni - Kluge, 1993

Aspidites Peters, 1876
Aspidiotes melanocephalus (Krefft, 1864)
Black-Headed Rock Snake, Matjdjun (Aboriginal), **Black-Headed Python**
Aspidites melanocephalus - Peters, 1876

Aspidiotes ramsayi (Macleay, 1882)
Ramsay's Python, Sand Python, Kuniya (Aboriginal), **Woma**
Aspidites collaris - Longman, 1913
Aspidites collaris - Wells & Wellington, 1984
Aspidites ramsayi - Peters, 1876

Liasis Gray, 1842
Liasis Mackloti (Dumeril & Bibron, 1844)
Australian Water Python, Brown Water Python, Swamp Python, Bolokko (Aboriginal), **Water Python**
Python timorensis - Müller, 1844
Python timoriensis - Müller, 1857
Liasis fuscus - W. Peters, 1873
Liasis corwallisius - Günther, 1879
Nardoa crassa - Macleay, 1885
Liasis fuscus - Stimson, 1969
Liasis mackloti - Smith, 1981b
Bothrochilus fuscus - Cogger et al., 1983
Lisalia fuscus - Wells & Wellington, 1984
Morelia fusca - Underwood & Stimson, 1990
Liasis mackloti - Kluge, 1993

Liasis olivacea (Gray, 1842)
Olive Rock Python, Mandjurdurrk (Aboriginal), **Olive Python**
Liasis olivaceus - Stimson, 1969
Lisalia olivaceus - Wells & Wellington, 1984
Lisalia barroni - Wells & Wellington, 1985
Morelia olivacea - Underwood & Stimson, 1990
Liasis olivaceus - Kluge, 1993

Leiopython Hubrecht, 1879
Liasis Albertisii (Peters & Doria, 1878)
D'Alberti's Rock Python, D'Albert's Python, **White-Lipped Python**
Leiopython gracilis - Hubrecht, 1879
Liasis albertisii - McDowell, 1975

Bothrochilus albertisii - Cogger et al., 1983
Lisalia albertisi - Wells & Wellington, 1984
Morelia albertisii - Stimson & Underwood, 1990
Leiopython albertisii - Kluge, 1993

Morelia Gray, 1842
Boa amethistina (Schneider, 1801)
Rock Python, Amethyst Rock Python, Amethyst Python, Amethystine Python, **Scrub Python**
Aspidopython Jakati - Meyer, 1874
Liasis duceboracensis - Günther, 1879
Hypaspistes dipsadides - Ogilby, 1891
Liasis clarki - Barbour, 1914
Liasis amethistinus kinghorni - Stull, 1933
Liasis amethistinus - Stimson, 1969
Python amethistinus - McDowell, 1975
Australiasis amethistinus - Wells & Wellington, 1984
Australiasis kinghorni - Wells & Wellington, 1984
Morelia amethistina - Underwood & Stimson, 1990
Morelia amethistina - Kluge, 1993

Python bredli (Gow, 1981)
Carpet Python, Carpet Snake, Bredl's Python, Bredl's Carpet Python, Centralian Python, Centralian Carpet Python, **Central Carpet Python**
Morelia bredli - Underwood & Stimson, 1990
Morelia spilota - Kluge, 1993

Python carinata (Smith, 1981a)
Rough-Scale Python, **Rough-Scaled Python**
Morelia carinata - Underwood & Stimson, 1990
Morelia carinata - Kluge, 1993

Python oenpelliensis (Gow, 1977)
Oenpelli Rock Python, Northern Rock Python, Narawan (Aboriginal), **Oenpelli Python**

Australiasis oenpelliensis - Wells & Wellington, 1984
Nyctophylopython oenpelliensis - Wells & Wellington, 1985
Morelia oenpelliensis - Underwood & Stimson, 1990
Morelia oenpelliensis - Kluge, 1993

Coluber spilotus (Lacepede, 1804)
Carpet Snake, **Carpet Python**; **Subspecific Forms:** Diamond Snake, **Diamond Python**; Atherton Tablelands Carpet Python, Rainforest Carpet Python, **Jungle Carpet Python**; Western Australian Carpet Python, South-Western Carpet Python, **Southwestern Carpet Python**; Common Carpet Python, Brisbane Carpet Python, Queensland Carpet Python, Coastal Carpet Python, **Eastern Carpet Python**; Murray-Darling River System Carpet Python, Murray Valley Carpet Python, Victorian Carpet Python; Inland Carpet Python, **Inland Rivers Carpet Python**; Northern Territory-Western Australia Carpet Python, Top-End Carpet Python, Northwestern Carpet Python, **Northern Carpet Python**.
Coluber Arges - Linnaeus, 1758
Python punctatus - Merrem, 1820
Python Peronii - Wagler, 1828
Morelia variegata - Gray, 1842
Morelia argus - Stimson, 1969
Python spilotus - McDowell, 1975
Morelia cheynei - Wells & Wellington, 1984
Morelia imbricata - Wells & Wellington, 1984
Morelia mcdowelli - Wells & Wellington, 1984
Morelia metcalfei - Wells & Wellington, 1985
Morelia spilota - Underwood & Stimson, 1990
Morelia spilota - Kluge, 1993

Python viridis (Schlegel, 1872)
Green Tree Python; **Green Python**
Chondropython azureus - Meyer, 1874
Chondropython pulcher - Sauvage, 1878
Chondropython viridis - Stimson, 1969
Morelia viridis - Underwood & Stimson, 1990
Morelia viridis - Kluge, 1993

Maps of Australia

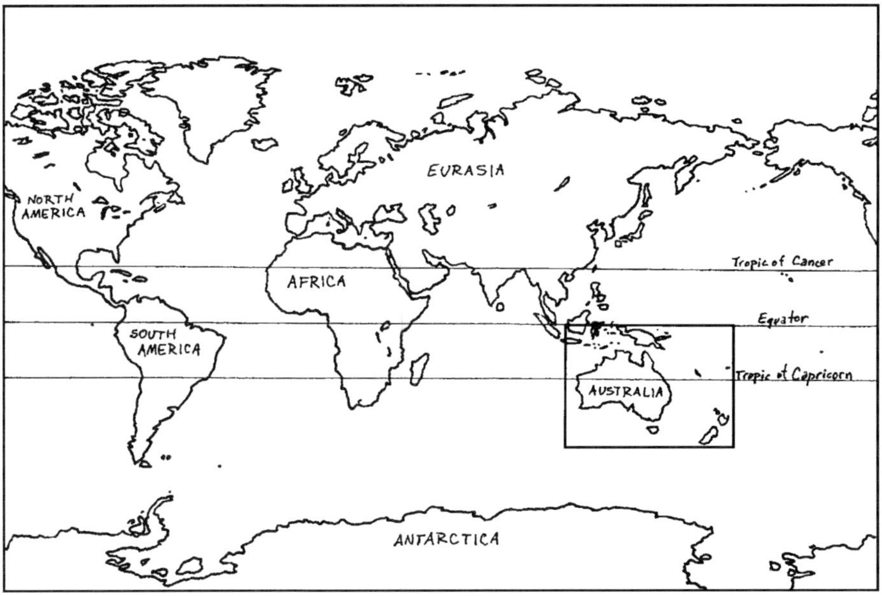

Figure 1. Australia, the only country that is a continent, lies entirely within the southern hemisphere, between the Indian and Pacific oceans. The name Australia is derived from the Latin word *australis*, which translates to southern.

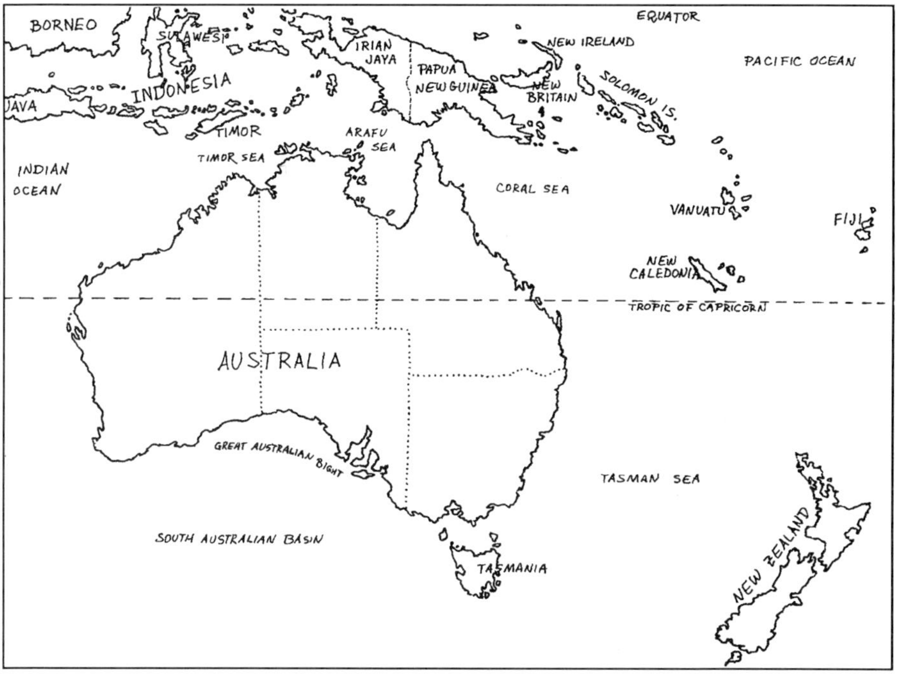

Figure 2. Australasia includes Australia, Indonesia, New Guinea, New Zealand, and other nearby islands. It is within this region that over 80% (32 of 40) of the presently recognized pythons occur, with 22 taxa occurring in Australia and 15 in Indonesia.

Figure 3. A political map of Australia that contains many of the localities mentioned in this book.

Rainfall (inches per year)

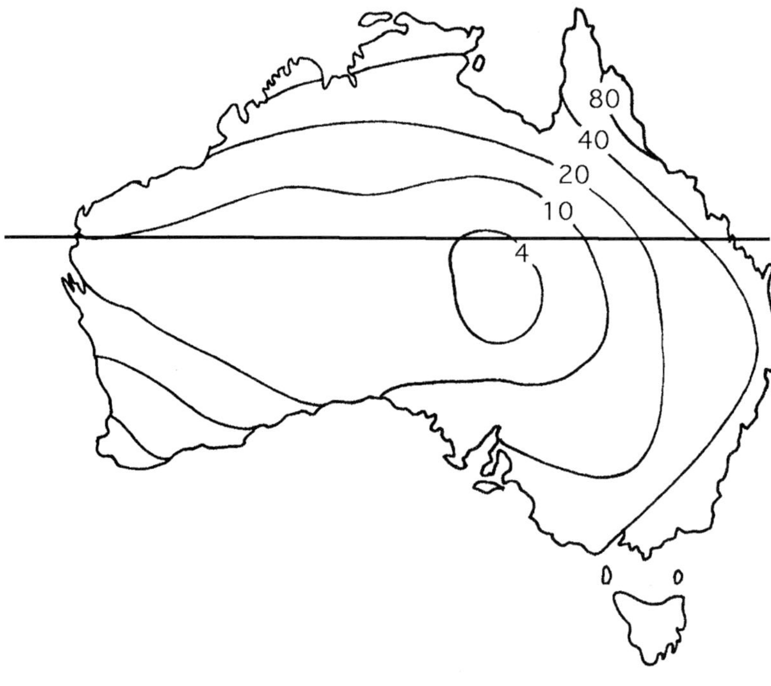

Figure 4. A map that shows Australia's average yearly precipitation in inches per year. Nearly all precipitation in Australia is rain, as snow only falls in highlands of the extreme southeast regions of the continent.

Vegetation

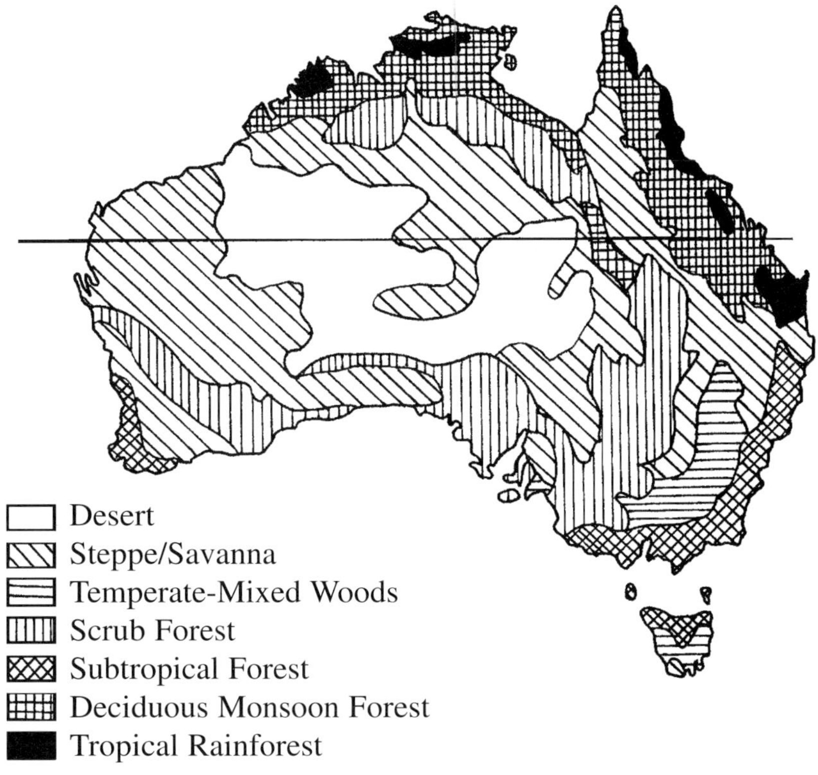

Figure 5. Vegetation in Australia is influenced by the frequency and amount of precipitation, which varies widely across the continent. Two native plants dominate the landscape. Eucalypts (*Eucalyptus*), also called gum trees, gums, or malees, are prevalent throughout areas of moisture, and are Australia's most common tree (~500 species). Acacias (*Leguminosae*), known as wattles, are common shrubs found throughout the dry regions of the continent. About 700 species are known, with the myall and the mulga being especially widespread.

Chapter 2

Ecology

Hiding and Basking

The ecology of pythons, as in other snakes, centers on hiding, basking, foraging, and reproducing. Most pythons are generally nocturnal in warm weather, but some taxa such as the Diamond Python are primarily diurnal (Shine, 1992). With the exception of searching for basking sites or mates, most pythons are relatively inactive during the day, preferring to rest in shelter (Fig. 8). Natural shelters vary depending on habitat and species, but commonly include hollow logs, rock crevices, rodent burrows, spinifex bushes (*Triodia*), termite mounds, thick foliage, and tree branches. The type of shelter depends on the habitats of the populations of python in question. For example, the terrestrial Olive Python often shelters in large rock outcroppings, and the semi-arboreal Central Carpet Python selects the hollow branches of the gum tree (*Eucalyptus*). Concealed from predators and extreme temperatures, pythons can digest a recent meal or rest.

Reptiles are ectotherms and thus unable to generate significant body heat through their own metabolic processes. The exception is that brooding female pythons of certain species can elevate their body temperature through thermal shivering (e.g., Ellis & Chappell, 1987).

Reptiles have evolved various mechanisms to utilize the environment to gain warmth. They have accomplished this through thermoregulatory behavior, which can be of three forms: concealed thermoregulation, partial basking, and complete basking.

Concealed thermoregulation, perhaps the most elementary type of thermoregulatory behavior in ectotherms, is accomplished within an animal's retreat. By selecting particular areas within its retreat, a python is able to utilize thermal gradients. As an area closer to the surface of a burrow or crevice is typically warmer during the day, particularly if exposed to direct sunlight, pythons are able to select from a wide range of temperatures while remaining concealed.

Termite mounds, rock outcroppings, and rock crevices are common sites for a python to conduct concealed thermoregulation. Not only do these sites provide a wide range of temperature variation, but also protection from predators (Figs. 13, 19). In some cases, these sites are inhabited by an abundance of prey items. For example, Spotted Pythons from northern New South Wales are difficult to encounter in the field, for they seldom leave the large rock formations and outcroppings in which they shelter. Rock formations provide them with numerous thermoregulatory opportunities and protection from predators. Prey items such as lizards, especially Spotted Dtellas (*Gehyra punctata*) and Tree Dtellas (*Gehyra variegata*), are abundant. Thus, the pythons only leave these refugia to search for water or mates.

When a python is concealed, this may not provide a desired range of temperatures; thus, a snake will emerge from its retreat and bask in the sun. Basking in snakes can be of two general forms. The first and most typical is when an individual rests in a protected area with only its head exposed to absorb sunlight. This method, termed partial basking, is advantageous to the snake, since it allows it to remain protected. Certainly, with only the head exposed, detection by predators would be more difficult. Partial basking is frequently observed in captivity. Specimens will often coil in their hidebox with only their head extended, absorbing heat from an incandescent light source situated above their hidebox. Under some situations extension of the head allows the snake to eliminate heat, as the retreat may be the warmest place available.

Another form of basking, termed complete basking, occurs when a snake leaves its shelter. Snakes use complete basking most frequently on cooler days, sometimes after initially warming up by partial basking. This method of basking has the disadvantage of exposing the snake to a potential predator, but the benefit is that the snake is able to warm itself more quickly by exposing a greater surface area to the sun (Fig. 43). Complete basking produces a notable increase in the body temperature and metabolic rate of pythons. Maximum body temperatures in free-ranging pythons are not well documented, although G. Bedford (*pers. comm.*) considers the preferred body temperature for active Water Pythons to be around 30°C. Repeated observations by Cogger & Holmes (1960) demonstrated that an Eastern Carpet Python which had basked the previous day, was between 1-18°F warmer the following morning, a considerable increase. Slip & Shine (1988a) discovered that adult Diamond Pythons will exhibit complete basking until elevating their body temperatures 28-30°C, and subsequently coil up tightly in a slightly cooler area. Coiling tightly enables the snake to retain sufficient heat to strike quickly and accurately at passing prey, even during cool nights. Slip & Shine (1988a) also found that juvenile Diamond Pythons are mainly diurnal hunters. Small specimens do not retain body heat well and lose heat quickly after sunset. Thus, as a consequence of thermal constraints, young Diamond Pythons generally prey on small diurnal lizards, such as the Copper-Tailed Skink (*Ctenotus taeniolatus*).

As a possible mechanism to absorb heat more effectively during the day, the Rough-Scaled Python and the Oenpelli Python undergo diel color changes (Charles & Wilson, 1985; Shine, 1991; Barker & Barker, 1994b). Two specimens of the Rough-Scaled Python were noted by Barker & Barker (1994b) to become noticeably paler in coloration at night, with less contrasting patterns. During daylight hours, the specimens were observed to have dorsal patterns of fawny red-brown and mottled cream. During daylight hours, Oenpelli Pythons are at times a drab chocolate brown with a barely visible dorsal pattern, but at night become light greyish-silver with a clearly visible dorsal pattern (Charles & Wilson, 1985; Shine, 1991)(Fig. 53).

Various species of *Antaresia*, *Liasis*, and other *Morelia* undergo less dramatic diel color shifts. Many specimens appear slightly lighter in coloration at night. Seasonal color changes also occur. The Woma, for example, in some areas darkens during the cooler months of winter (Shine, 1991). This is considered unusual in pythons, but it is well documented among the larger Australian elapids, such as the Inland Taipan (*Oxyuranus microlepidotus*) and Western Brown Snake (*Pseudonaja nuchalis*) (Shine, 1991; Hoser, 1993). Another type of color variation is biogeographic, and occurs in the Western Stimson's Python, the Southwestern Carpet Python, and the Inland Rivers Carpet Python. Perhaps as a result of a cooler microhabitat, specimens are darker in the south and southwestern parts of their range. Specimens of the Western Stimson's Python and the Southwestern Carpet Python from forests and offshore islands, where the climate is generally cool, are much darker than specimens from northern sand plains (Figs. 9, 64, 80).

Foraging Tactics

Pythons generally emerge from their daytime retreats at dusk to hunt for food or search for mates. Some species, such as the Diamond Python, are ambush hunters and remain in one spot for several days until an animal passes within striking distance (Slip & Shine, 1988d).

Pythons locate and subdue prey by using one or a combination of four senses: olfaction, vomeronasal, visual, and thermal. Pythons often detect potential prey by chemosensation, sometimes tracking a targeted animal's movements for hundreds of meters. When a python closes in on its prey, it utilizes both its vision and heat-sensitive organs, which are located within the labial pits.

Labial pits are small depressions in the labial scales which border the mouth of most boids. Within these pits, certain cells respond to infrared radiation, and they are sensitive to minute temperature variations. These organs aid pythons in hunting and, possibly in egg incubation. With the exception of the genus *Aspidites*, all pythons have labial pits (Figs. 10, 11). The absence of labial pits in *Aspidites* may be explained by their unique semi-fossorial behavior, as both members of the genus have been documented to excavate soil under rocks and open crevices with their head (Fyfe & Harvey, 1981; O'Brien & Naylor, 1987; Templeton, 1990). Perhaps digging and fossorial behaviors precluded the evolution of labial pits, since plugging of the pits would render them less useful for thermal detection.

When a python determines that a prey item is within striking distance, the snake will strike, bite, and immediately coil. Its many needle-sharp, recurved teeth hold the animal in place while constriction begins. Contrary to popular thinking, death is not caused primarily by suffocation. Constriction, which is the result of the python's muscles tightening around prey, causes death by circulatory arrest, followed by rapid tissue hypoxia of vital organ systems (Hardy, 1994). When the snake has killed its prey, eating commences. Depending on the size of the prey, feeding may take from a several minutes to several hours. Larger prey is generally consumed head first and it appears that many pythons will rub their snouts against these items, possibly searching for the head. The direction of hair, feathers, or scales of the prey, reveals the position of the head.

In response to environmental conditions, some pythons have evolved specialized hunting and feeding behaviors. The Black-Headed Python and the Woma frequently enter rodent burrows in search of food. In subterranean environments, these snakes utilize an alternative method of killing prey, by pressing it against the walls of the burrow (Fyfe & Harvey, 1981). This method is a modified form of constriction, and is commonly observed in colubrid snakes of the genera *Elaphe* and *Lampropeltis*; it is particularly useful in subduing multiple prey items. During this process the snake can be injured through bites and scratches. These interactions can be quite severe, as wild Black-Headed Pythons and Womas often show extensive scarring.

Another foraging tactic used by snakes is caudal luring, a type of mimicry which involves the attraction of potential prey through a series of writhing tail movements that have the general appearance of a worm or larva (Schuett et. al., 1984; Strimple, 1992). The prey, such as a small lizard, frog, bird, or mammal, is attracted to the snake's tail because of mistaken identity. When the prey nears the tail or attacks it, the snake, which has been lying perfectly still, strikes out and captures its prey. Caudal luring occurs in a wide variety of snakes, including numerous pitvipers, dwarf boas (*Tropidophis* spp.) and death adders (*Acanthophis* spp.), and two species of python, the Green Python and the Woma. In captivity, the Green Python is known to frequently use caudal luring as a foraging technique (Murphy et. al., 1978). Observations in captivity show that neonates and juveniles of *M. viridis* caudal lure at a greater frequency than adults. This is most likely due to the ontogenetic shift in their diet, with juveniles feeding primarily on lizards and adults on birds and mammals (Henderson, 1993).

The Woma is the only terrestrial python known to caudal lure. This behavior was first reported by Fyfe & Harvey (1981), and was found to be similar to that of the Death Adder. Womas from the majority of their range have been observed to caudal lure, but the frequency of this behavior is not well documented. It may vary not only between individual specimens, but possibly between regional populations.[10] An unusually high number of Womas with stub tails are found in nature, and such specimens are well represented in museum collections (Kluge, 1993). Although predator attacks can not be ruled out, it has been suspected that these injuries are the result of prospective prey. Since caudal luring may not be present in all Woma populations, a particularly interesting explanation was provided by Casey Lazik (*pers. comm.*). Apparently, Womas in certain areas are prone to having groups (N = 20) of ticks attach to their tail tips. The degree of tick infestation makes shedding on the tail tip difficult, and several wild specimens were observed to have accumulated layers of shed skin on the tail. A buildup of layers of shed skin, in addition to damage caused by the ticks, presumably would cause extensive scarring, tissue damage, and the loss of the tail tip.

Diet

All snakes are carnivorous. The feeding ecology of Australian pythons remains poorly understood, but certain generalizations can be made. Smaller species, such as members of the *A. childreni* group, and hatchlings of all Australian pythons are largely lizard feeders, primarily preying on geckos (*Diplodactylus, Gehyra,* and *Oedura*) and small skinks (*Carlia, Ctenotus, Eremiascincus, Lampropholis,* and *Morethia*) (Figs. 6, 12, 13).

Mammals and birds make up the bulk of the diet of Australian pythons (Figs. 6, 14). This trend holds true for most of the larger species, except for the Black-Headed Python and Woma. These pythons prey mainly on other reptiles, including goannas and bearded dragons.

One of the more unusual prey items are bats. Bent-winged bats (*Miniopturus* spp.) and fruit bats (*Pteropus* spp.) are major components of the diet for several Australian pythons. Smaller species, such as Northern Brown Pythons and Spotted Pythons, frequently feed on Bent-Winged Bats (*M. schreibersii*), a species approximately the size of a laboratory mouse.

Larger pythons, such as the Water Python, Olive Python, Scrub Python, Oenpelli Python, Eastern Carpet Python, and Northern Carpet Python feed on flying foxes, which can have a wing span of over 39 inches (>1 meter) in length and a body as large as an adult laboratory rat. For several months of the year, the Oenpelli Python frequents the tops of fruit-producing trees of the Arnhem Land escarpment to prey on the flying foxes and native pigeons that rest in the branches (Fig. 8).[11]

Throughout Australia, pythons are often responsible for consuming family pets. The Northern Brown Python, Spotted Python, Stimson's Python, Water Python, Scrub Python, and Carpet Pythons are known to devour small rodents, rabbits, birds, and occasionally cats. Pythons frequently prey upon the birds in aviaries and are often unable to squeeze out of the cages after they have ingested the prey. During the summer months, Queensland rodent breeders report both

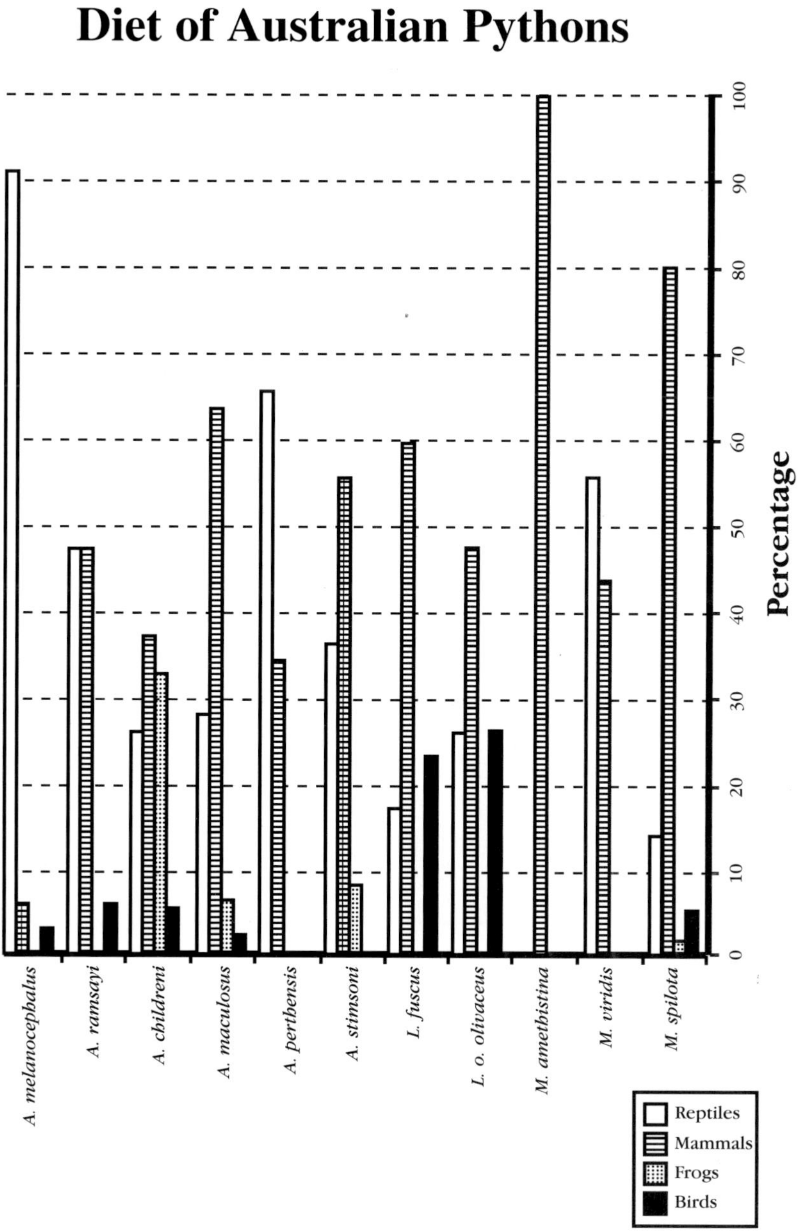

Figure 6. A histogram of the diet of Australian pythons based on data collected through museum dissections and field observations by Shine (1991, p. 208), which shows general predatory trends in eleven species. In captivity, pythons will feed on other prey items (e.g., *M. amethistina* on birds).

Spotted and Scrub pythons are sometimes discovered inside rodent breeding sheds. Three large Scrub Pythons were captured in a Queensland poultry breeder's chicken enclosure in a single evening.[2] In some cases pythons will consume any animal of suitable size, sometimes leading to peculiar and interesting situations. Lethbridge (1993) describes encountering a Scrub Python of over 13 feet (4.0 m) in length constricting an Echidna (*Tachyglossus aculeatus*) near Bramwell Holding, Cape York Peninsula, Queensland. In spite of the quills, the Scrub Python devoured this interesting prey item.

Predators and Defense

Pythons are preyed upon by a large variety of animals, including various raptors, large lizards, other snakes, crocodiles, dingoes, the introduced European Fox (*Vulpes vulpes*), and the domestic cat. None of these predators, however, feed on pythons exclusively.

Smaller species of Australian pythons are more frequently preyed upon by large birds and elapids. In contrast, the larger species have relatively few predators once they reach sexual maturity. Adult Olive Pythons, Scrub Pythons, and Oenpelli Pythons are not considered to have many natural predators except for crocodiles (*Crocodylus*) and dingoes.

Humans obviously kill pythons for various reasons. For thousands of years, Aborigines have used goannas (*Varanus*), File Snakes (*Acrochordus arafurae*), and pythons as a source of meat. Humans also kill pythons for sport, or out of fear and ignorance. And with the invention of the automobile, a host of reptiles have become frequent victims of vehicular traffic (Fig. 74). In the Barkly Tablelands, Northern Territory, Black-Headed Pythons are sometimes encountered at night while crossing the Barkly Highway by truck drivers transporting cattle. At truck stops, these pythons are often secretly placed in other trucks as pranks.[17]

Agricultural development often impacts reptile populations through the destruction of their habitat. This has occurred in southwestern Australia, where populations of the Woma and the

Southwestern Carpet Python apparently have declined as a direct result of agriculture. Agricultural development also indirectly affects python populations by bringing them into greater contact with humans. For example, pythons and other animals fall victim to introduced predators, such as feral cats and the European Fox. The Black-Headed Python and the Woma do not resemble typical pythons and frequently are mistaken for venomous snakes and killed. The Water Python, Olive Python, and Pilbara Olive Python, due to their large size and uniform coloration, often are confused with the highly venomous Brown Snake (*Pseudonaja nuchalis* and *Pseudonaja textilis*) and King Brown Snake (*Pseudechis australis*) and killed (Pearson, 1993).[22]

Like most snakes, pythons are secretive creatures and this provides their principal defense. They rely on their coloration and ability to function silently. When cornered, however, all pythons are capable of defense, but when possible opt for escape (Fig. 15). In an attempt to appear more threatening, juvenile Water Pythons, and occasionally juvenile Black-Headed Pythons and Olive Pythons, often assume a vertical posture and gape.[22] Neonate Black-Headed Pythons have been noted to flatten their head in a manner similar to that of the Brown and King Brown snakes (Barker, 1984; Barker & Barker, 1994b).[22] When aggressive behavior fails, one species, the Stimson's Python, is known to coil into a tight ball with the head concealed in the middle (Hoser, 1989; Ehmann, 1993). All pythons can swim, and some, including the Water Python, Olive Python, and Pilbara Olive Python take to the water when disturbed. Ehmann (1993) noteed that several pythons, including the Water Python, Olive Python, Pilbara Olive Python, Scrub Python, and Carpet Python, are able to remain submerged in water for 20-45 minutes.

If held by the head, pythons will constrict until they are released. In all but the smaller species, this often results in a painful bite. Pythons also are capable of regurgitation and excretion as defensive tactics. Scrub Pythons, in particular, are infamous for these behaviors.[22]

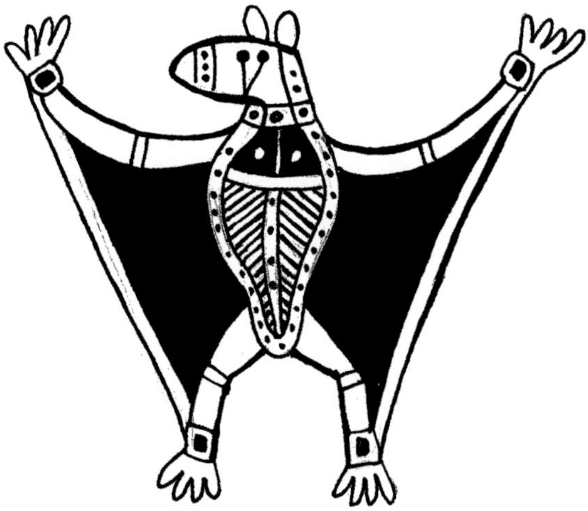

Color Plates

Figure 7. This is Ubrir Rock, Kakadu National Park, Northern Territory. Kakadu National Park, one of the best areas to observe Australia's python diversity, is home to six species (Black-Headed, Northern Brown, Water, Olive, Oenpelli, and Northern Carpet pythons).

William B. Love

Figure 8. An Oenpelli Python, or Narawan, to the Aborigines of Arnhem Land, photographed at Ngarradj Wade Djobkeng, Kakadu National Park, Northern Territory. This specimen was discovered near a sandstone ridge and under a large White-Apple Tree (*Sysygium forte*). Photographer Ian Morris, a park ranger at Kakadu, reports that Oenpelli Pythons often are encountered coiled under these trees, as their prolific flowers and blossoms attract lorikeets (*Trichoglossus*) during the day and colonies of flying foxes (*Pteropus*) at night. This specimen had eaten a large flying fox the previous evening.

Figure 9. Many pythons are darker in color in the southwestern parts of their range, probably as a result of cooler temperatures. A Western Stimson's Python from Millendon, 15 miles (24 km) NE of Perth, Western Australia, is an example of this trend.

Ian Morris

Brad Maryan

Figure 10. A Woma from Barkly Highway, west of the Roadhouse, Barkly Tableland, Northern Territory, found in an area of red sand with spinifex and *Acacia*. Labial pits, which are small, heat sensitive depressions in the labial scales, are absent in the genus *Aspidites*.

Figure 11. An Oenpelli Python from Nabarlek, Arnhem Land, Northern Territory, shows well-defined labial pits. Labial pits aid in hunting and possibly in defense and egg incubation.

Matthew Milgate

Greg Harold

Figure 12. A juvenile Spotted Python from Esk, Queensland, found eating a Stone Gecko (*Diplodactylus vittatus*). Geckos are common prey of Australia's pythons.

Figure 13. Northern Brown Pythons search the overhangs of rocky caverns and galleries for geckos, small bats, and birds in the Kimberleys, Western Australia. This specimen, from Kununurra, emerged from an abandoned nest of a Fairy Martin (*Cecropis ariel*). The nests are built of mud and cover the walls and overhangs of many caves, where they serve as refugia for a variety of animals, including juvenile Rock Monitors (*Varanus glebopalma*) and Spotted Tree Monitors (*V. scalaris*).

Steve Wilson

Steve Wilson

Figure 14. A Scrub Python from near Malanda, Atherton Tableland, Queensland, that has devoured a large wallaby, the Red-Legged Pademelon (*Thylogale stigmatica*). The snake was found perched on a sapling tree nearly 5 feet (1.5 m) above the ground along a swift moving stream.

C. & D. Frith

Figure 15. Pythons are capable of defending themselves vigorously when provoked. This juvenile Diamond Python from the Hornsby Region, north of Sydney, New South Wales, was found coiled tightly under a rock in an area of weathered sandstone.

Matthew Milgate

Figure 16. (**Top**) This locality is near Sydney, New South Wales. (**Bottom**) Terrible bushfires swept through these woodlands and surrounding areas in January, 1994. These fires had a disastrous effect on the local herpetofauna. Diamond Pythons probably fared poorly, especially females incubating their eggs.

Greg Harold

Corey Blanc

Figure 31. A Pygmy Python from Meka Homestead, Murchison Goldfields, Western Australia, located in a granite outcrop. Pygmy Pythons are most frequently encountered on warm, humid evenings, either crossing roads or prowling around termitaria in rocky hummock grasslands.

Figure 32. This locality is Circular Pool, Hamersley Range National Park, Pilbara, Western Australia. Here, the Pygmy, Black-Headed, and Pilbara Olive pythons inhabit the ironstone rock faces and caves in deep gorges. The rugged Hamersleys stretch nearly 300 miles (ca. 500 km) across the arid Pilbara and are known for their deep brown, rich red, and violet rocks.

Greg Harold

Brad Maryan

Figure 33. A juvenile western Stimson's Python from Strelly Homestead, 37 miles (60 km) E of Port Hedland, Pilbara, Western Australia. Immature Western Stimson's Pythons from the Pilbara region typically have pale interspaces between distinct transverse blotches.

Figure 34. The tumbled boulder formations with surrounding spinifex plains at Harding River Dam, near Roebourne, Pilbara, Western Australia, are habitat for the Western Stimson's, Pygmy and Black-Headed pythons.

Robert Browne-Cooper

Brad Maryan

Figure 35. A Western Stimson's Python from Old Rainy Rocks, Eastern Goldfields, Western Australia found under a flat slab of granite.

Figure 36. Granite rock hills with scattered spinifex at Cherratta Homestead, near Karratha, Pilbara, Western Australia are habitat for the Western Stimson's, Black-Headed, Pygmy, and Pilbara Olive pythons.

Greg Harold

Brad Maryan

Figure 37. A juvenile Eastern Stimson's Python found under a rock slab in a house garden at the Alice Springs Telegraph Station, 2.5 miles (4 km) N of Alice Springs Town Centre, Northern Territory.

Figure 38. The Eastern Stimson's Python and Woma are found throughout the rocky spinifex plains surrounding the Olgas, near Ayers Rock, Uluru National Park, Northern Territory.

Greg Fyfe

William B. Love

Figure 39. A reputed hybrid between the Northern Brown Python and Eastern Stimson's Python found in an area of black soil clay 6-9 miles (10-15 km) S of Brunette Creek, on the Tablelands Highway of Brunette Downs Station, Barkly Tableland, Northern Territory.

Figure 40. The dry, cracking, black soil clay plains of the Barkly Tableland, Northern Territory, are where the Northern Brown and Eastern Stimson's pythons frequently hybridize. Most specimens look like intermediates; some bear a closer resemblance to the Northern Brown Python, while others to the Eastern Stimson's Python.

Greg Fyfe

Greg Fyfe

Figure 41. This juvenile Water Python was encountered on an overgrown river floodplain near the junction of Route 36 and the Mary River, Northern Territory.

William B. Love

Figure 42. (**Top**) A view from the levee at Fogg Dam, Northern Territory looking west into a billabong, and (**Bottom**) looking east into the dry floodplain. In the evening, Water Pythons frequently cross the levee apparently searching for the Floodplain Rat (*Rattus lutreolus*).

William B. Love

Figure 43. An Olive Python from the King Leopold Ranges, Kimberleys, Western Australia. This specimen was found basking beside several large sandstone boulders. Specimens from the Kimberleys often show extensive scarring, more so than other species from Australia. This may result from males combatting, or perhaps their choice of prey.

Figure 44. Sandstone outcrops alongside a billabong in the King Leopold Ranges, Kimberleys, Western Australia, provide habitat for the Olive and Northern Brown pythons. At night, Olive Pythons frequent intermittent pools formed when rivers or small streams cease to flow.

Corey Blanc

Corey Blanc

Figure 45. A Pilbara Olive Python from Indee Station, Pilbara, Western Australia displays the dusty, red-brown coloration sometimes occurring in this taxon. This python frequents the vicinity of permanent waterpools in the harsh country along the Fortescue River.

Figure 46. A juvenile Pilbara Olive Python from the De Grey River, Pilbara, Western Australia. This specimen was found with a spotlight nearly 10 feet (3.0 m) up a gum tree above a nearly dried water pool. It had just eaten three finch nestlings, and had down feathers clinging to the side of

Ron Johnstone

Corey Blanc

Figure 47. Banded ironstone rock faces and deep waterpools such as Nangilgardy Pool, near Paraburdoo, south of the Hamersley Ranges, Pilbara, Western Australia are habitat for the Pilbara Olive Python. Black-Headed, Pygmy, and Western Stimson's pythons inhabit the nearby

Greg Harold

Figure 48. A Scrub Python from Cairns, Queensland. On cooler days, specimens actively search the rainforest canopy for patches of sunlight in which to bask.

Figure 49. A Scrub Python from Iron Range National Park, Cape York Peninsula, Queensland. Unlike Carpet Pythons, male Scrub and Olive pythons are usually larger in size. From late July to early September, male Scrub Pythons often show large wounds resulting from intense combat.

Steve Wilson

Steve Wilson

Figure 50. The Scrub Python inhabits the lowland rainforest of Dunk Island, Queensland. Pictured is a strangler fig (*Ficus*) which is native to tropical forests in Queensland, New Guinea, and southeast Asia. Beginning life as a seed high in another tree, the strangler fig drops its roots after germination. Over time, the tangled root masses grow to enclose the parent tree, while the fig grows upwards toward the light of the canopy. The parent tree eventually dies from a lack of nutrients and light, and as it rots, a hollow is left in the center. These hollows are important shelters to many creatures of the rainforest.

C. & D. Frith

igure 51. A Rough-Scaled Python from 25 miles 40 km) NE of Mt. Trafalgar, Kimberleys, Western ustralia. This specimen was discovered in a patch f vine forest at the edge of a sandstone cliff.

Figure 52. Mitchell Falls, west Kimberleys, Western Australia, is the type locality for the Rough-Scaled Python, and suitable habitat for the Northern Brown, Olive, and Northern Carpet pythons. Although the Rough-Scaled Python was once thought to primarily inhabit the sandstone rock formations of this region, it may actually prefer monsoon rainforests.

Ron Johnstone

Corey Blanc

Figure 53. An Oenpelli Python from the small mining settlement of Nabarlek, Arnhem Land, Northern Territory, displaying its light grey-silver nocturnal colors, which render it cryptic among the escarpment's sandstone outcroppings.

Figure 54. Oenpelli and Northern Brown pythons are found in the caves, crevices, and fractured sandstone formations along the East Alligator River, Kakadu National Park, Northern Territory. Also found in this region are Black-Headed, Water, Olive, and Northern Carpet pythons. Aboriginal rock paintings at nearby Ubrir Rock (Fig. 7) are a popular tourist attraction. These paintings depict a variety of subjects and styles; some works are several thousand years old, while others were painted early this century. Motifs similar to Aboriginal rock paintings have been found as far away in place and time as Upper Paleolithic Europe.

Greg Harold

Greg Miles

Figure 55. This Diamond Python was found in the rafters of a barn in a semi-forested suburb of Sydney, New South Wales.

Figure 56. Diamond Pythons inhabit the sandstone and sclerophyll forest of this Sydney valley. This taxon is found in a temperate area, where it tends to inhabit rocky gullies and woodlands. It was in this region that David Slip and Dr. Richard Shine studied Diamond Pythons using radio-telemetry, a first for any boid snake.

Matthew Milgate

Matthew Milgate

Figure 59. A Jungle Carpet Python photographed in the rainforests of Paluma, just north of Townsville, Queensland, at an elevation of 3000 feet (914 m). This specimen measured slightly over 3 feet (1 m).

Figure 60. A Jungle Carpet Python from near Boulders, Babinda, Queensland, found crossing a road at night in an area of dense, closed canopy rainforest adjacent to a rushing stream.

C. & D.Frith

Lyall Naylor

Figure 61. A Jungle Carpet Python from Thompson Creek, Cape Tribulation National Park, southern Cape York Peninsula, Queensland. These unusual Jungle Carpet Pythons inhabit the closed canopy rainforest and restricted waterways of Noah Creek, Mile Creek, and Thompson Creek. Typically they have black and grey-white stripes, but in some cases, their bands are black and ivory white.

Figure 62. Jungle Carpet and Scrub pythons inhabit the upland rainforest with Orania palms of Mt. Lewis, Atherton Tablelands, Queensland. Throughout the forested slopes of nearby Millstream Falls National Park, Lake Barrine National Park, Bellenden Ker National Park, and Palmerston national parks, Duck-Billed Platypuses (*Ornithorhynchus anatinus*) can be seen swimming in calm pools alongside turbulent rivers and waterfalls. During the wet season from December to March, this area receives about 138 inches (3500 mm) of rainfall, and is one of the wettest areas in Australia.

Lyall Naylor

C.& D. Frith

Figure 63. A Southwestern Carpet Python from Two Rocks, about 37 miles (60 km) N of Perth, Western Australia. This subadult specimen was rescued from a suburban home. Pythons sometimes are released by farmers near grain sheds to control rodent populations.

Figure 64. This is Sullivan Rock, Darling Ranges near Perth, Western Australia. A typical granite outcrop with exfoliated slabs and *Eucalyptus* forest provides habitat for the Southwestern Carpet and the Western Stimson's pythons.

Brad Maryan

Brad Maryan

Figure 65. An Eastern Carpet Python from Iron Range National Park, Cape York Peninsula, Queensland. This specimen was found in a riverine vine forest along a creek line between dense rainforest patches.

Figure 66. The rainforest of Maiala National Park, Mt. Glorious, about 25 miles (40 km) NW of Brisbane, Queensland, is home to the Eastern Carpet Python. Further north, on the Cape York Peninsula west of the Great Dividing Range, populations of the Eastern Carpet Python are found under different conditions. In these areas, seasonally dry tree-lined streams and lagoons are the habitat of this wide-

C. & D. Frith

Steve Wilson

Figure 67. This Eastern Carpet Python from Malanda, Atherton Tablelands, Queensland, is an example of the larger, darker Carpet Pythons also found throughout the range of the Jungle Carpet Python. It was found at an elevation of 2600 feet (792 m).

C.& D. Frith

Figure 68. An Inland Rivers Carpet Python from 19 miles (30 km) N of Currawinya National Park, southwest Queensland, from an area of arid shrubland adjacent to a *Eucalyptus* lined creek. This specimen is typical of those found along the northern extensions of the Murray Valley and Lake Eyre drainage basins.

Figure 69. Shallow meandering waterways and seasonally dry *Eucalyptus* lined river flats along the Paroo River in southwest Queensland, are habitat for the Inland Rivers Carpet Python.

Steve Wilson

Steve Wilson

Figure 70. A Northern Carpet Python found in an area of lowland forest along the South Alligator River, Kakadu National Park, Northern Territory.

Figure 71. The lush grassy woods and trees of Litchfield National Park, Northern Territory, are primary habitat for the Northern Carpet, Northern Brown, and, possibly, the Black-Headed and the Olive pythons. Photographed on the onset of the rainy season in early December, the monsoon wetlands and grassy woodlands of Litchfield, Kakadu, and many northern regions are rejuvenated every year during the four

Greg Miles

William B. Love

Figure 72. A Green Python from the tropical rainforest of West Claudie Creek, Iron Range National Park, Cape York Peninsula, Queensland. This specimen was found resting on a large log about 3 feet (1 m) above the ground.

Figure 73. Dense lowland rainforest at West Claudie Creek, Iron Range National Park, Cape York Peninsula, Queensland, is the habitat of the Green Python. In nearby localities at Iron Range, Spotted, Scrub, and Eastern Carpet pythons have been found.

C. & D. Frith

C. & D. Frith

Figure 74. Vehicles are responsible for the decimation of many reptilian populations. Here, a Black-Headed Python from about 12 miles (20 km) SW of Eighty Mile Beach Turnoff, Western Australia, is a victim of road traffic.

Casey Lazik

Chapter 3

Reproductive Biology

Phenology of Reproduction

Australian pythons are seasonal breeders in nature (Field, 1990). The reproductive behavior and physiology of pythons are controlled by endogenous rythms, as well as environmental variations in temperature, humidity, photoperiod, barometric pressure, and rainfall.

In Australia, free-ranging pythons mate during the winter (June-August), eggs are deposited in the spring (September-November), and hatching occurs in the late spring and summer (December-February) (Field, 1990). Australia is in the southern hemisphere (Figs. 1, 2), thus, the seasons are in reverse to the northern latitudes. These reproductive trends correspond to the winter mating, spring ovipositing, and summer hatching most frequently reported by American and European breeders (Ross & Marzec, 1990; Barker & Barker, 1994b).

In many areas of Australia, nocturnal temperatures are lower than breeders provide their animals in captivity. Night-time temperatures of 65-70°F (8-21°C) are generally provided to captive pythons during the breeding season, but temperatures often are lower in nature. For

Corey Blanc

Corey Blanc

Figure 45. A Pilbara Olive Python from Indee Station, Pilbara, Western Australia displays the dusty, red-brown coloration sometimes occurring in this taxon. This python frequents the vicinity of permanent waterpools in the harsh country along the Fortescue River.

Figure 46. A juvenile Pilbara Olive Python from the De Grey River, Pilbara, Western Australia. This specimen was found with a spotlight nearly 10 feet (3.0 m) up a gum tree above a nearly dried water pool. It had just eaten three finch nestlings, and had down feathers clinging to the side of its mouth.

Ron Johnstone

Corey Blanc

Figure 47. Banded ironstone rock faces and deep waterpools such as Nangilgardy Pool, near Paraburdoo, south of the Hamersley Ranges, Pilbara, Western Australia are habitat for the Pilbara Olive Python. Black-Headed, Pygmy, and Western Stimson's pythons inhabit the nearby

Greg Harold

Figure 48. A Scrub Python from Cairns, Queensland. On cooler days, specimens actively search the rainforest canopy for patches of sunlight in which to bask.

Figure 49. A Scrub Python from Iron Range National Park, Cape York Peninsula, Queensland. Unlike Carpet Pythons, male Scrub and Olive pythons are usually larger in size. From late July to early September, male Scrub Pythons often show large wounds resulting from intense combat.

Steve Wilson

Steve Wilson

Figure 50. The Scrub Python inhabits the lowland rainforest of Dunk Island, Queensland. Pictured is a strangler fig (*Ficus*) which is native to tropical forests in Queensland, New Guinea, and southeast Asia. Beginning life as a seed high in another tree, the strangler fig drops its roots after germination. Over time, the tangled root masses grow to enclose the parent tree, while the fig grows upwards toward the light of the canopy. The parent tree eventually dies from a lack of nutrients and light, and as it rots, a hollow is left in the center. These hollows are important shelters to many creatures of the rainforest.

C. & D. Frith

Figure 51. A Rough-Scaled Python from 25 miles (40 km) NE of Mt. Trafalgar, Kimberleys, Western Australia. This specimen was discovered in a patch of vine forest at the edge of a sandstone cliff.

Figure 52. Mitchell Falls, west Kimberleys, Western Australia, is the type locality for the Rough-Scaled Python, and suitable habitat for the Northern Brown, Olive, and Northern Carpet pythons. Although the Rough-Scaled Python was once thought to primarily inhabit the sandstone rock formations of this region, it may actually prefer monsoon rainforests.

Ron Johnstone

Corey Blanc

Figure 53. An Oenpelli Python from the small mining settlement of Nabarlek, Arnhem Land, Northern Territory, displaying its light grey-silver nocturnal colors, which render it cryptic among the escarpment's sandstone outcroppings.

Figure 54. Oenpelli and Northern Brown pythons are found in the caves, crevices, and fractured sandstone formations along the East Alligator River, Kakadu National Park, Northern Territory. Also found in this region are Black-Headed, Water, Olive, and Northern Carpet pythons. Aboriginal rock paintings at nearby Ubrir Rock (Fig. 7) are a popular tourist attraction. These paintings depict a variety of subjects and styles; some works are several thousand years old, while others were painted early this century. Motifs similar to Aboriginal rock paintings have been found as far away in place and time as Upper Paleolithic Europe.

Greg Harold

Greg Miles

Figure 55. This Diamond Python was found in the rafters of a barn in a semi-forested suburb of Sydney, New South Wales.

Figure 56. Diamond Pythons inhabit the sandstone and sclerophyll forest of this Sydney valley. This taxon is found in a temperate area, where it tends to inhabit rocky gullies and woodlands. It was in this region that David Slip and Dr. Richard Shine studied Diamond Pythons using radio-telemetry, a first for any boid snake.

Matthew Milgate

Matthew Milgate

Figure 59. A Jungle Carpet Python photographed in the rainforests of Paluma, just north of Townsville, Queensland, at an elevation of 3000 feet (914 m). This specimen measured slightly over 3 feet (1 m).

Figure 60. A Jungle Carpet Python from near Boulders, Babinda, Queensland, found crossing a road at night in an area of dense, closed canopy rainforest adjacent to a rushing stream.

C. & D.Frith

Lyall Naylor

Figure 61. A Jungle Carpet Python from Thompson Creek, Cape Tribulation National Park, southern Cape York Peninsula, Queensland. These unusual Jungle Carpet Pythons inhabit the closed canopy rainforest and restricted waterways of Noah Creek, Mile Creek, and Thompson Creek. Typically they have black and grey-white stripes, but in some cases, their bands are black and ivory white.

Figure 62. Jungle Carpet and Scrub pythons inhabit the upland rainforest with Orania palms of Mt. Lewis, Atherton Tablelands, Queensland. Throughout the forested slopes of nearby Millstream Falls National Park, Lake Barrine National Park, Bellenden Ker National Park, and Palmerston national parks, Duck-Billed Platypuses (*Ornithorhynchus anatinus*) can be seen swimming in calm pools alongside turbulent rivers and waterfalls. During the wet season from December to March, this area receives about 138 inches (3500 mm) of rainfall, and is one of the wettest areas in Australia.

example, in central Australia, the habitats of the Black-Headed Python, Woma, Eastern Stimson's Python, and Central Carpet Python may experience temperatures below 32°F (0°C). In areas of New South Wales inhabited by the Diamond Python, temperatures will routinely fall to 40°F (4°C). Even in the rainforests of Northern Queensland, temperatures can fall as low as 40°F (4°C) (Szokolay, 1983). It is unlikely that, because of concealment, pythons are exposed to these low temperatures for long, if at all.

In captivity, wild-caught Green Pythons from New Guinea have been observed copulating during storms and low pressure fronts, indicating that reproductive behavior in wild pythons may not be dependent solely on declining nocturnal temperatures (Ross & Marzec, 1990). Because factors influencing the reproductive cycles of wild pythons remain poorly understood, we should not dismiss unusual hypotheses regarding factors such as lunar cycles.

Courtship and Copulation

Male pythons are known to travel great distances, sometimes up to 500 m/day in search of females (Shine, 1991). During this time, males tend to cease foraging.

Courtship is variable in duration, and is characterized by the male slowly following the female, constantly flicking his tongue. During courtship, which is the pre-copulatory phase of reproduction, males frequently prod females with small spurs located on both sides of the cloaca. The spurs are the remnants of hind limbs, and may project and vibrate in an attempt to stimulate the female into moving forward to give the male better alignment with her cloaca. This process may be quite forceful, often causing visible scratches and hisses from a female (Ross & Marzec, 1990). In some species, the scratching of the male's spurs against the female is audible.

At some point the male mounts the female, positions his tail and cloaca near hers, and, if she is receptive, inserts a hemipenis (snakes have a bi-lobed reproductive organ) into her cloaca. A receptive female may elevate her tail and expose her cloaca. This behavior is termed cloacal gaping. In some species, particularly the Water Python and Olive Python, blood spotting of the substrate has been observed along with cloacal gaping (Ross & Marzec, 1990; Barker & Barker, 1994b).

Male Aggregation and Combat

Python courtship and copulation usually takes place between a single male and female. In some instances, however, several males, in detecting a female nearby, may arrive to court a single female. This behavior, known as male aggregation, has been observed in other snakes, such as the Red-Sided Garter Snake (*Thamnophis sirtalis parietalis*), in which dozens of males may court a single female. In garter snakes, however, males do not engage in direct fights. In pythons, low numbers of males aggregate with females. Male aggregation has been observed in the Black-Headed Python, Scrub Python, Diamond Python, and Carpet Pythons.[22]

During male aggregation, combat often determines dominance (Fig. 17). In the Diamond Python male combat has not been observed, but in the closely related Carpet Pythons, male combat includes vicious biting. Normally, only dominant individuals, which are those who win contests, gain priority of access to females and subsequently copulate with them (Schuett, 1997). Barker et. al (1979) showed that captive male Indian Pythons (*Python molurus*) formed a linear hierarchy marked by unchanging social position. Bedford & Sullivan (1993) observed male social behavior in captive Northern Brown Pythons, and found that these hierarchies were stable.

Unfortunately, there are no detailed reports on long-term social structure in male Australian pythons in the wild, although it is well known that most species engage in combat (Shine, 1994). Taxa such as the Black-Headed Python, Woma, Northern Brown Python, Spotted Python, Eastern Stimson's Python, Western Stimson's Python, Water Python, Olive Python, Pilbara Olive Python, Scrub Python,

Central Carpet Python, Jungle Carpet Python, Eastern Carpet Python, Northern Carpet Python, and Green Python have been reported to show male combat (Barker & Barker, 1994b).[9,10,22] For the taxa not mentioned here, this is probably a better indication that both wild and captive observations are lacking, than a judgment as to whether male combat actually occurs.

Research on the social behavior of captive Indian Pythons can serve as a model for interpreting male combat among Australian pythons. As described by Barker et. al (1979), male combat in the Indian Python begins with a recognition-investigation phase in which two sexually-excited male pythons approach each other, setting the stage for the ascent-alignment-orientation phase in which one male, generally the dominant animal, will attempt to orient himself along the dorsum of the second male. This phase soon leads to actual wrestling, in which the combatants will repeatedly entwine around, each snake trying to maintain its head above the other's. This activity, referred to as topping, is the third phase of male combat. Finally, in the submission phase, the individual establishing dominance may elevate its head and anterior trunk above the head of the subordinate male for varying lengths of time.

Among Australian pythons, combat can become very intense with dominant males sometimes injuring subordinate ones by biting. In captivity, male combat can lead to injury or death, since subordinate males are unable to escape (Ross & Marzec, 1990.)

Ovulation, Gestation, and Unusual Behavior of Gravid Pythons

Testicular and ovarian growth in pythons occurs during the mating season (Barker & Barker 1994b, 1995b). During this time, females will ovulate. Once mating has occurred, the follicles are released into the body cavity, now referred to as ova, and enter the oviducts, where fertilization occurs. Females show posterior bending (midbody swelling) at this time, as the ova are pushed forward into the oviducts. This swelling generally lasts from 12-24 hours. In the Black-Headed Python and the Olive Python, a massive midbody swelling lasts up to four days.[22]

After fertilization, embryos develop in the female for 2-2.5 months. In nature, this period can never be measured exactly because females may mate multiple times, perhaps to assure reproductive success. Wild female pythons also may be subjected to fluctuating temperatures, which may affect the length of gestation. The concept of "degree days" as discussed by Ross & Marzec (1990) states that the temperature in degrees per twenty-four day multiplied by the length of gestation in days yields a figure often helpful in determining temperature ranges and thermal requirements for gravid pythons. In the field, where constant conditions are not present, it would not be surprising that periods of gestation were slightly longer than in captivity.

In captivity, where specimens are separated after a certain date and environmental conditions regulated, relatively accurate gestation dates can be determined. Boos (1979) reported that the gestation period for a female Eastern Carpet Python at Sydney's Taronga Park Zoo was 82 days. Bush (1988) reported the period between the first observed mating to oviposition in a female Southwestern Carpet Python as 80 days. During gestation, two unusual behaviors are sometimes observed in female pythons, i.e., dramatic changes in coloration and inverted postures of the body.

Captive female pythons have been observed to undergo dramatic color shifts during the breeding season and period of egg production (Charles, 1988; Hingley & Hemmings, 1993). Charles (1988) described a gravid Scrub Python that became almost black dorsally. Darkening, which can be a sign of pregnancy, also may increase heat absorption during gestation (Ross & Marzec, 1990). In contrast to typical observations of darkening in gravid females, Hingley & Hemmings (1993) reported that a female Eastern Carpet Python changed from a dull, blue-black with light greenish bands to a dark brown with golden bands and an accompanying iridescence previously undisplayed. Further research appears to be necessary to understand both the ecological and physiological significance of color change in female pythons.

Another unusual behavior is observed in gestating pythons is referred to as the inverted posture (Fig. 18). This position, in which the abdominal area is rotated upward to about 180°, exposes the python's ventral scales. As a result, the female appears to be lying on her back. In a related behavior, gravid Green Pythons often hang their lower regions of their body down from the branch from which they are perched. The reasons for these postures are unclear, but several observations can be included.

Captive female pythons begin to exhibit inverting behavior several weeks after mating, and will coil in the inverted posture, particularly when basking under a spotlight. With the exception of the Black-Headed Python and the Woma, inverting is discontinued following oviposition. In these pythons, both males and females position themselves in a manner similar to the inverted posture as a normal way of coiling. This posture, therefore, does not always indicate a gravid condition.

Little is known about the habits of gravid pythons in nature, hence very few free-ranging pythons have been seen in the inverted posture. In one case, a Northern Brown Python was observed just before sunset in early November, 1968, basking in the inverted posture on the bank of Alexander Creek, which joins into the Leichhardt River near the Gulf of Carpentaria in Queensland.[14] A gravid female Northern Brown Python was observed basking in an inverted posture in mid-afternoon in December, 1974, along the bank of the Gregory River in the Northern Territory's Gulf of Carpentaria.[4] Also, a Water Python was sighted near Townsville, in Giru, Queensland, in the inverted posture. The specimen was found in the morning in direct sunlight, resting upon a large, rotting wood pile in a swampy lagoon near the water's edge.[14]

Oviposition, Maternal Incubation, and Thermoregulatory Behavior

Several days to several weeks before oviposition, female pythons may shed their skin and become restless, presumably in search of a suitable nesting site. In captivity this particular shed is referred to as the pre-egg laying shed and is used to gauge when oviposition is imminent. In most taxa, oviposition generally occurs about two to three weeks after the pre-egg laying shed.

In order to support embryonic development, the female python generally maintains her eggs between 86-91°F (30-32°C). As described by Walsh (1979), it has been hypothesized that there are at least four methods by which incubating pythons are able to regulate clutch temperature and humidity. These methods, which may be used singly or concomitantly, include the selection of the incubation site, the position of the female on her eggs, thermal shivering, and basking.

When looking for an oviposition site, a female python chooses a refugium that is secure, possibly a familiar retreat. As described earlier, female pythons seem to choose from a wide range of retreats having various thermal and humidity conditions. For example, Water Pythons from Fogg Dam in the Northern Territory have been observed to incubate egg clutches in abandoned goanna burrows within the dam walls (Shine, 1991). The Pygmy Python and Stimson's Python have been observed to incubate their eggs in the base of termite mounds. Termite mounds often are used as oviposition sites by a number of reptile species, such as goannas, due to stable temperature and humidity levels maintained inside (Hoser, 1993) (Fig. 19).

The Diamond Python is apparently unique among pythons in that gravid females actively build nests in which to deposit their eggs. Shine (1992) describes that nests are made in sandy areas with leaf litter, generally 20-70 meters from a creek. A distinct mound is formed, generally in leaf litter and under a tree or bush.

Charles *et al.* (1985) reported on two wild Eastern Carpet Pythons from Brisbane, Queensland that were discovered while brooding eggs. One female was found under building materials on 2 January, and observed for a week. During that time she frequently basked in the sun and maintained her clutch of 21 eggs at 86°F (30°C). The second female was not noted to leave her eggs. The eggs were laid on a grass clipping mound, and the snake remained coiled around them even during rainfall. Through muscular contractions, she maintained the temperature of her egg clutch at a constant 86°F (30°C). About one month later, she was observed to lay only seven eggs, but bulges from retained eggs remained evident. The female was removed from her eggs, and those were artificially incubated. The retained eggs were removed surgically.

When a suitable site is located, a female python coils tightly, and within the next several days begins to lay her eggs. Python eggs generally adhere to one another shortly after oviposition to form a solid mass. Following oviposition, the female will incubate her clutch, and when necessary, will thermoregulate. Incubation is accomplished by the female coiling around her eggs and maintaining them at a suitable temperature and humidity through the use of different coiling positions. During suitable temperature and humidity levels, a female python will coil loosely around her eggs. When temperatures are at a suboptimum level, the female will coil tightly around the eggs and slowly twitch, and through muscular contraction will elevate her body temperature and warm the clutch. In some species, this slow twitching occurs at short intervals, about every 15-30 seconds.

Likewise, if temperatures should rise above the optimum level, a female will sometimes shift the entire clutch vertically and lift the eggs off the ground with her coils. Williams (1992a) and Foster (1993) reported that female Spotted Pythons will shift their egg clutches horizontally. In one instance, a female moved her clutch about 400 mm across her cage (Williams, 1992a). This remarkable behavior has been observed among captive specimens. Foster (1993) noted that free-ranging Northern Brown Pythons have been found incubating their eggs beneath sheets of paperbark trees (*Melaleuca*) in Northern Queensland's Gulf of Carpentaria country. Perhaps the female shifts her eggs as the sun moves and heats different parts of the paperbark sheet.

During thermoregulation of eggs, a female python sometimes will position her head inside the coils to the extent that it is not observable. This behavior may be a form of temperature monitoring, since the female python's heat-sensitive labial pits may function like a thermometer, indicating when warming or cooling methods should be used (G. Schuett, *pers. comm.*). According to Hoser (1989) and Barker & Barker (1994b), a python's labial pits are sensitive to temperature variations as little as 0.1°C or less.

Remarkably, measurements taken from the coils of some female pythons show that they can keep their eggs almost 6°C higher than the ambient temperature (Hoser, 1989). When several brooding females are using the same shelter, as has been noted in the Ball Python (*Python regius*), greater ambient temperatures can be generated (Greer, 1994).

It is likely that all Australian pythons maternally incubate their eggs. Thermal shivering during incubation, however, probably varies intra- and interspecifically (Barker & Barker, 1994b; Ross & Marzec, 1990). Taxa observed to shiver in captivity include the Black-Headed Python, Northern Brown Python, Spotted Python, Water Python, Olive Python, Scrub Python, Diamond Python, Central Carpet Python, Jungle Carpet Python, Eastern Carpet Python, and Green Python (Barker & Barker, 1994b).[9,10,22] It is currently unknown if the Woma, Pygmy Python, Stimson's Python, Pilbara Olive Python, Rough-Scale Python, Oenpelli Python, Southwestern Carpet Python, Inland Rivers Carpet Python, and Northern Carpet Python show thermal shivering, as reproductive data for these taxa are scant or lacking altogether. Future observations may show the majority of the above species to exhibit thermal shivering during maternal incubation.

Some Australian pythons may not be capable of thermal shivering, as other pythons such as the Ball Python (*Python regius*) and the

African Rock Python (*Python sebae*) do not exhibit thermal shivering (Ellis & Chappell, 1987; Ross & Marzec, 1990; Barker & Barker, 1994a). Those species probably rely on basking for warming their eggs. Incubating pythons, both in the field and in captivity, have been observed to leave their clutches to bask after a cool night.

A female python will remain coiled around her eggs, guarding them from predators and maintaining proper temperatures and proper humidity levels until they hatch. Wild female pythons likely defend their egg clutches vigorously against predators, such as goannas (Shine, 1991). Captive observations have shown that many female pythons, which have settled around their eggs, will strongly defend them against removal or inspection by striking repeatedly at intruders. In contrast, other females remain placid during egg removal or inspection.

There are no conclusive data to support the view that wild female pythons maintain proper humidity levels for their clutches of eggs, but it is logical to assume that the site selected for incubation has consistent and high levels. Captive female pythons engaged in maternal incubation have been observed to raise the humidity of their brooding chambers by soaking in water and returning to their eggs thoroughly wet (Ross & Marzec, 1990). Shine (1991) stated that captive female Water Pythons leave their clutches to drink and, upon returning, excrete fluids from their cloaca onto their eggs. Free-ranging pythons may use similar methods to maintain high humidity levels.

Thermal shivering and fasting often result in significant weight loss to the female during the protracted incubation period. Wild female pythons probably do not hunt while incubating their eggs, and thus, undergo aphagia for a period of three to five months. In free-ranging Diamond Pythons, this results in a loss of nearly half of their pre-oviposition body mass, which leads to a biennial or triennial female reproductive cycle (Shine, 1992). Except in favorable habitats with adequate food supplies, females probably are unable to replenish their fat bodies and weight reserves before the next breeding season.[19]

Hatching, Growth Rates, and Age at Maturity

In many species of Australian pythons, eggs hatch about 60 days after oviposition. Incubation time, however, can vary depending on environmental conditions and the species in question. For most taxa, this period is from 45-75 days. In the Black-Headed Python, Woma, Olive Python, and Oenpelli Python, incubation time, at least in captive specimens, is longer than in other species, sometimes up to 100 days (Ross & Marzec, 1990).

Hatching is a potentially stressful process for neonate pythons. It begins when the young python slits its egg with a tiny egg tooth, a short-lived structure on the rostrum that is specifically used to cut through the tough, leathery egg shell (Fig. 20). The young snake then slowly extends its head out of the egg, returning quickly back inside if it feels threatened. Over the next 24-48 hours, the egg yolk will be absorbed before the snake emerges from the egg completely. After hatching, neonates may move away from the clutch site, though some may remain in the general area for several days.

From several hours to about two weeks after birth, neonate pythons will shed their skin for the first time and along with it their egg tooth. In the only breeding record available for the Oenpelli Python, the first shed did not occur until the neonates were two months of age. The neonates in this captive group accepted small birds prior to their first shed, which is considered unusual for hatchling pythons (Krauss, 1992). In most hatchling pythons, searching for food begins shortly after the first shed.

In tropical areas of Australia, pythons hatch at the beginning of the wet season, which is usually in December. This coincides with the increased biological activity brought about by the rains (Field, 1990). Feeding generally commences quickly, as there is an abundance of small lizards, frogs, and mammals available for food. Like many reptiles, neonate pythons grow rapidly their first year.

Unfortunately, little is known on growth rates and age of maturity in free-ranging pythons. Shine (1991) provided data on the growth of Water Pythons at Fogg Dam in the Northern Territory. It appears that specimens from this area reach sexual maturity at 3.25 feet (100 cm) and one year of age. This is an interesting account, since captive specimens seldom mature and reproduce before three years of age (Ross & Marzec, 1990). Although captive pythons in general grow faster than their wild counterparts, in some microhabitats, like Fogg Dam, prey may be so plentiful that food consumption and growth rates exceed those of captivity.

In captivity, most Australian pythons become reproductively active between 3-4 four years of age, although there are exceptions. The Olive Python is physically mature at three years of age, but most captive breedings generally have occurred with specimens from 8-12 years of age.[22] One pair of Black-Headed Pythons, which were received as adults in 1980, were reported to copulate almost every year after 1982 before producing infertile eggs in 1990 and 13 viable eggs in 1995.[16] At the other end of the spectrum, species such as the Woma, Northern Brown Python, and Spotted Python can reproduce when they are 18-24 months of age.

Chapter 4

The Australian Pythons

Key to the Genera of Australian Pythons

1.	Labial scales with sensory pits present	2
	Labial scales without sensory pits present	*Aspidites*
2.	Small, irregular scales on head	*Morelia*
	Large, "shield-like" scales on head	3
3.	One or two loreal scales present	4
	More than two loreal scales present	*Antaresia*
4.	One pair of prefrontal scales	*Leiopython*
	More than one pair of prefrontal scales	*Liasis*

Genus *Aspidites*
Peters, 1867

Aspidites is a genus of terrestrial pythons that are endemic to Australia. There are two species, the Black-Headed Python and the Woma, which are identified from all other pythons by the absence of heat sensory labial pits.

Key to the Species of *Aspidites*

Head and neck black; numerous crossbands of brown, black, or red present on the body	*melanocephalus*
Head and neck yellow, brown, or olive; dorsum is brown, olive, or pale yellow	*ramsayi*

Black-Headed Python
Aspidites melanocephalus (Krefft, 1864)

The Black-Headed Python is found throughout the northern sections of Western Australia, the Northern Territory, and northern Queensland (Fig. 75). The species lives in a variety of habitats, such as sand ridges, rocky hill country, dry savanna, and tropical grassland (Figs. 7, 22, 24, 32, 34, 36, 47, 54, 71, 74). Several geographical color phases are known and are associated with specific habitats. One population has distinct bands and is light in contrast, and their reddish to black bands are mixed with white or cream ground color. It inhabits the red sandy ridges of the western edge of the Great Sandy Desert, from Port Hedland along Eighty Mile Beach, to Anna Plains, Western Australia (Figs. 21, 74). Specimens from this area tend to be smaller in size than from other populations, averaging about 5.0-5.5 feet (1.53-1.69 m) in length.

South from Anna Plains, in the rocky ranges of the Pilbara and Hamersley Ranges, Black-Headed Pythons are slightly larger and darker in coloration, and by all accounts I could obtain, are similar to the specimens from the savanna populations of the Kimberleys and Northern Territory. In tropical grassland populations of the Northern Territory, particularly throughout Litchfield and Kakadu National Parks, specimens have an overall yellowish-grey appearance.[22] Some specimens from coastal forest regions of Queensland have rusty brown to red bands and are very dark in overall coloration. Individuals in all of these areas are notably larger than western specimens, commonly reaching 9-10 feet (2.4-2.7 m). In general, specimens from Queensland are the largest, and

Black-Headed Pythons from the red ridges of the western Great Sandy Desert are striking in appearance, with bands of red, dark brown, or black interspersed by cream-white. Specimens average 5.0 -5.5 feet (1.53-1.69 m) in length.

Black- Headed Pythons from the coastal forests of Queensland have rusty brown to red bands and appear very dark in overall coloration. Adults average about 9-10 feet (2.77-3.08 m) in length, but specimens measuring over 12 feet (3.69 m) have been recorded.

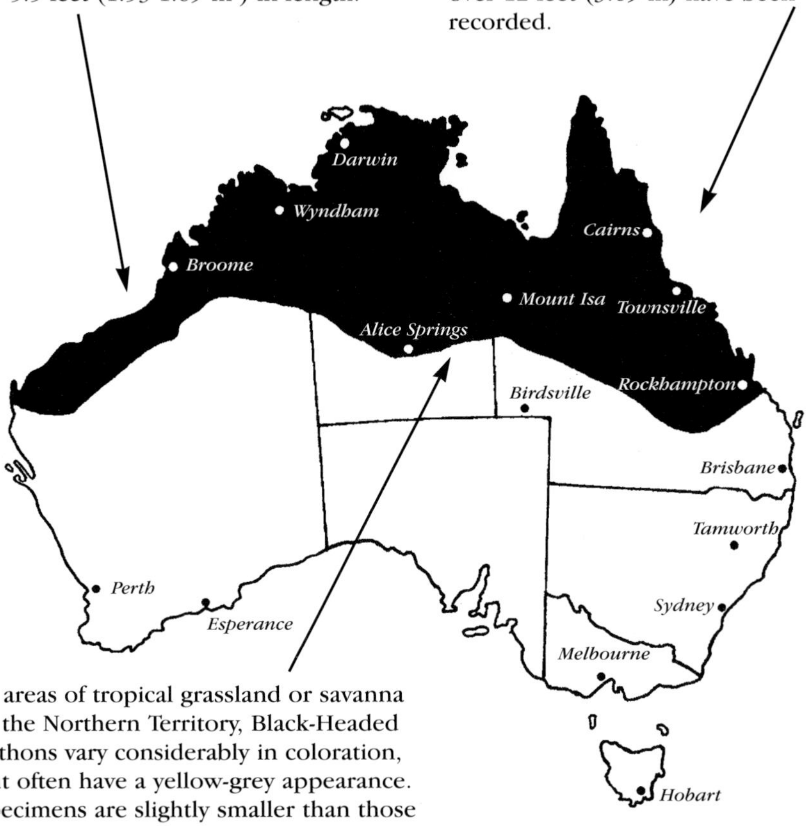

In areas of tropical grassland or savanna in the Northern Territory, Black-Headed Pythons vary considerably in coloration, but often have a yellow-grey appearance. Specimens are slightly smaller than those from Queensland, averaging 8-9 feet (2.46-2.77 m) in length.

Figure 75. Distribution of the Black-Headed Python.

one specimen measured thirteen feet (4.0 m).[22] This species is commonly encountered throughout its range. Specimens can be found coiled tightly, completely exposed, or alongside asphalt roads. Occasionally, they are found during the summer months crossing these asphalt roads, which are far too hot to transverse without shoes in broad daylight.

Black-Headed Pythons are nocturnal hunters and mainly prey on other reptiles (Fig. 6), such as goannas, bearded dragons, skinks, and other snakes, including highly venomous elapids. A freshly-killed specimen, measuring 5.5 feet (1.69 m) was found on the road about 33 miles (52.8 km) southwest of the Anna Plains Station Turnoff, Western Australia. When its stomach contents were examined, a Woma measuring about 3.5 feet (1.08 m) was found.[15]

The Black-Headed Python is not considered to be cannibalistic in captivity. During the breeding season, however, one specimen was reported to devour a cagemate. It is not known whether cannibalism occurs in nature (Charles & Field, 1985).

Females generally oviposit 4-14 eggs. Barker & Barker (1994b) record 20 clutches ranging from 6-18 eggs. Hatchlings measure 20-24 inches (50-61 cm). Although Barker (1984) reported that in his experience Black-Headed Pythons do not exhibit sexual behavior until five years of age, 10 clutches are known from specimens under three years.[9, 22]

Woma

Aspidites ramsayi (Macleay, 1882)

The Woma is found throughout the sandy regions of Western Australia, South Australia, northern New South Wales, southwestern Queensland, and the Northern Territory (Fig. 76). Throughout its range it exhibits several distinctive color forms. Whether or not taxonomic recognition is warranted certainly merits further investigation (Barker & Barker, 1994b).[10]

Specimens from northern Western Australia and the Northern Territory tend to have a cream to yellow background with orange-red to brown dorsal bands (Figs. 10, 23, 25). They average 4-5 feet (1.2-1.5 m)

Figure 76. Distribution of the Woma.

in length. Specimens from Queensland, referred to erroneously as *A. collaris* (Longman, 1913; Wells & Wellington, 1984), have an orange-yellow head and a grey-brown dorsolateral pattern. Their eyes are surrounded by blackish circles or bars, giving them the name "black-eyed" Woma. Specimens from this population exhibit an ontogenetic color change. As juveniles they have a brilliant orange-yellow head with a bold greyish banded dorsal appearance. As they mature this color fades, and adults have overall brown-yellow dorsal coloration and banding is less apparent. The blackish circles remain, but they are much less prominent. They seldom exceed 5 feet (1.5 m) in length. Another distinctive population occurs throughout the southern regions of the distribution. These include specimens from south Western Australia to specimens from the Simpson Desert, Northern Territory. Many individuals from these populations are drab brown, sometimes almost black to olive green in color, and quite large in size, reaching 8-9 feet (2.4-2.7 m) in length. Authorities generally regard Womas from the Simpson Desert as the largest. Many adults are over 7 feet (2.2 m) in length, and a few reach 10 feet in length (3.0 m).[10]

The population of Womas found in south-Western Australia is considered rare, possibly endangered, and their conservation status may be in need of urgent attention. A combination of agricultural development, feral predators, and mortality resulting from human activity apparently has caused the population to decline. Records from the Western Australian Museum suggest that they were once fairly abundant (Smith, 1981a).

Opinions differ as to whether populations of *A. ramsayi* should be considered taxonomically distinct on account of their disjunct distribution. It is likely, however, that most populations are not isolated. Greg Fyfe (*pers. comm.*) who has had extensive field experience with *A. ramsayi* writes:

> The southwest population of southern Womas, regarded as isolated by Smith (1981a), Wilson & Knowles (1988), and Hoser (1989) is not actually isolated as indicated in these works. Smith (1981a) discounts a sight record that "bridges" the gap between the south-western population of southern Womas and the Womas of the Nullarbor plain area because the record came

> from a non-sandy area. However, among many authorities, this record is considered valid, as other Womas have been found in non-sandy areas throughout Australia. For example, here in Alice Springs, we have had two separate road-kills in an area of mulga and heavy soils, not typical habitat for Womas, but the snakes were still there.

The Woma is sympatric with the Black-Headed Python throughout parts of their range, specifically the Pilbara region of Western Australia (Figs. 24, 74). Field observations indicate that the Woma tends to inhabit flatter, sandy areas, as opposed to the rocky terrain preferred by the Black-Headed Python (Hoser, 1992) (Figs. 24, 26, 38). Specimens from the Northern Territory and Western Australia are often found on sandy dunes, frequently in and nearby rodent and rabbit burrows. Womas apparently will enter the rabbit burrows in search of nestling rabbits, which in these areas constitute a substantial portion of their diet.

Python breeders in Australia have observed matings in Womas in every month of the year, but eggs are laid seasonally.[10] Clutch size appears to vary in size with specimens from various localities. For example, large southern Womas can lay a clutch over 20 eggs, while Womas from other areas generally lay clutches of 7-10 eggs. Barker & Barker (1994b) record 17 clutches ranging from 5-19 eggs. Gow (1989) indicates that Joe Bredl hatched 28 eggs from a female collected in Moomba, South Australia. Neonates are about 18 inches (40-50 cm) in length.

Genus *Antaresia*

Wells & Wellington, 1984

Antaresia is a genus of terrestrial pythons unique to Australia. The genus *Antaresia* differs from other python genera by the presence of regular shield-like head scales.

Key to the Species of *Antaresia*

1.	Dorsal pattern tends to disappear	2
	Persistent dorsal pattern	3
2.	31-35 midbody scale rows	*perthensis*

	36-46 midbody scale rows	*childreni*
3.	Smooth-edged, elongate blotches or transverse bars	4
	Ragged-edged, irregular chocolate blotches, often joining to form a broken zig-zag stripe on the dorsum	*maculosus*
4.	260-302 ventral scales	*stimson stimsoni*
	243-284 ventral scales	*stimsoni orientalis*

Northern Brown Python
Antaresia childreni (Gray, 1842)

This species is found throughout northern Western Australia, the Northern Territory, and Queensland (Fig. 77). This species generally does not surpass 39 inches (1.0 m) in length, and it is one of Australia's smallest pythons.

The Northern Brown Python is variable in dorsal color, ranging from grey to various shades of red (Figs. 13, 27). Adults have little dorsal pattern and are sometimes completely unmarked, whereas juveniles are boldly marked. These bold patterns fade as specimens mature. In some adults, small, irregular, smooth-edged, purple-brown blotches barely contrast with the ground color (Smith, 1985). Throughout much of its range, including the Kimberleys of Western Australia, Katherine, Northern Territory, and Queensland's Mt. Isa, unmarked and patterned specimens occur in the same population.

The Northern Brown Python has adapted to a variety of habitats, from dry savanna and grassland to coastal forest (Figs. 7, 44, 52, 54, & 71). Specimens are often encountered in rocky hills with spinifex bushes, especially near permanent water.[13] Like other members of the genus, the Northern Brown Python frequently crosses roads. Rocky outcroppings and crevices are prime areas of shelter, but other retreats, such as termite mounds and spinifex bushes, are used. The Northern Brown Python feeds on small lizards, mainly geckos and skinks, rodents, birds, and occasionally frogs (Hoser, 1981b) (Fig. 6). Shine & Slip (1990) found this species to have the most varied diet of any Australian python. One specimen also was observed to pursue, subdue,

Northern Brown Pythons in the Kimberleys of Western Australia, Katherine, Northern Territory, and Mount Isa, Queensland may be patternless or patterned.

Figure 77. Distribution of the Northern Brown Python.

and eat a non-venomous Keelback Snake (*Tropidonophis mairii*).[4] Prior to this observation, ophiophagy was unreported in this species.

Clutch size in the Northern Brown Python is approximately 10 eggs, which generally are oviposited in December. Barker & Barker (1994b) record 24 clutches ranging between 7-20 eggs. The neonates emerge after 45-60 days of incubation and are approximately 10 inches (25 cm) at birth.

Spotted Python

Antaresia maculosus (Peters, 1873)

The Spotted Python, one of several species formerly classified as *A. childreni*, is found along coastal Queensland from the Islands of the Torres Strait southward nearly 300 km into New South Wales, and on numerous islands off the Queensland coast (Smith, 1985) (Fig. 78).

This species is distinguished from the Northern Brown Python and Stimson's Python by a dorsal pattern consisting of chocolate-brown, ragged-edged blotches which tend to join anteriorly and posteriorly, and often form a broken zig-zag stripe on the dorsum (Smith, 1985) (Figs. 12, 29). Barker & Barker (1994b; 1995a) report on a population of Spotted Pythons from the central eastern coast of the Cape York Peninsula, in which specimens attain a large size at maturity, have prominent markings, and yellow ventral scales. This morph of the Spotted Python is highly sought after by python fanciers. With the exception of this population, the variation in populations of Spotted Pythons from the Cape York Peninsula is poorly documented. Speckled and patternless morphs have hatched from clutches in the United States, but these specimens are not reported in the wild or in Australian collections.[22] The Spotted Python is relatively small, reaching maturity at about 3 feet (1 m) in length. The largest specimens are from the areas surrounding Tully, northern Queensland, where some individuals are known to reach 6.5 feet (2 m) in length.[9]

This python tends to inhabit rocky hills and ridges and also is found in coastal forests and savannas, generally in close proximity to ephemeral

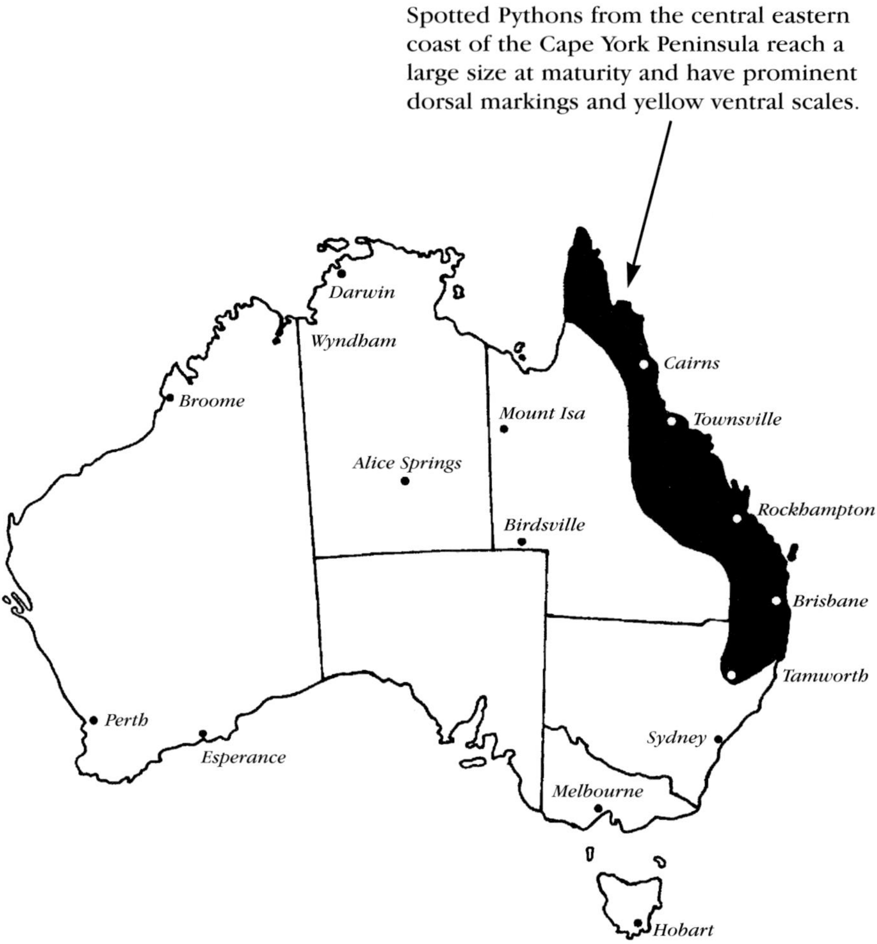

Figure 78. Distribution of the Spotted Python.

streams (Figs. 30, 73). This species tends to search for prey at night and stay concealed within crevices by day. When rock formations are unavailable, they often seek refuge within hollow logs and in leaf litter.[13] Prey is mostly small rodents, lizards, and insectivorous bats (Fig. 6). Hoser (1989) describes numerous caves near Mt. Etna, in close proximity to Rockhampton, Queensland where a colony of several thousand bent-winged and ghost bats (*Miniopturus* spp.) were observed to be the prey for several hundred Spotted Pythons.

The Spotted Python is commonly bred in captivity. Clutch size is generally 10-16 eggs. Barker & Barker (1994b) record 11 clutches, each with 4-18 eggs. Based on captive observations, Barker & Barker (1994b) speculate on a south-to-north increase in clutch size, egg size, and hatchling size. Neonate Spotted Pythons are about 12 inches (30 cm) in length. In captivity they grow quickly, and some reach maturity and reproduce at 18 months of age.

Pygmy Python

Antaresia perthensis (Stull, 1932)

The Pygmy Python is found in the Pilbara region of Western Australia (Fig. 79). It is the world's smallest python, reaching maturity at 18-24 inches (45-62 cm) in length. Pygmy Pythons are various shades of brick-red with no visible dorsal pattern or having a slight pattern (Fig. 31).

This species inhabits the rocky ranges and spinifex grasslands of the Pilbara (Figs. 22, 32, 34, 36, 47). Individuals frequently seek shelter in large termite mounds or ant-hills. Within the tunnels of the termite mounds, the Pygmy Python feeds on small skinks, geckos, and marsupial mice (*Antechinus* spp.) (Hoser, 1989) (Fig. 6). Cogger (1992) noted that the late F. J. Mitchell found individuals inhabiting the tunnels of a giant termitaria of *Eutermes triodiae*. Apparently, they prey on the Pilbara Dtella (*Gehyra pilbara*), which also inhabits the tunnels of the termitaria. The Pygmy Python shelters among rocky outcroppings or spinifex bushes when mounds are not available.[13]

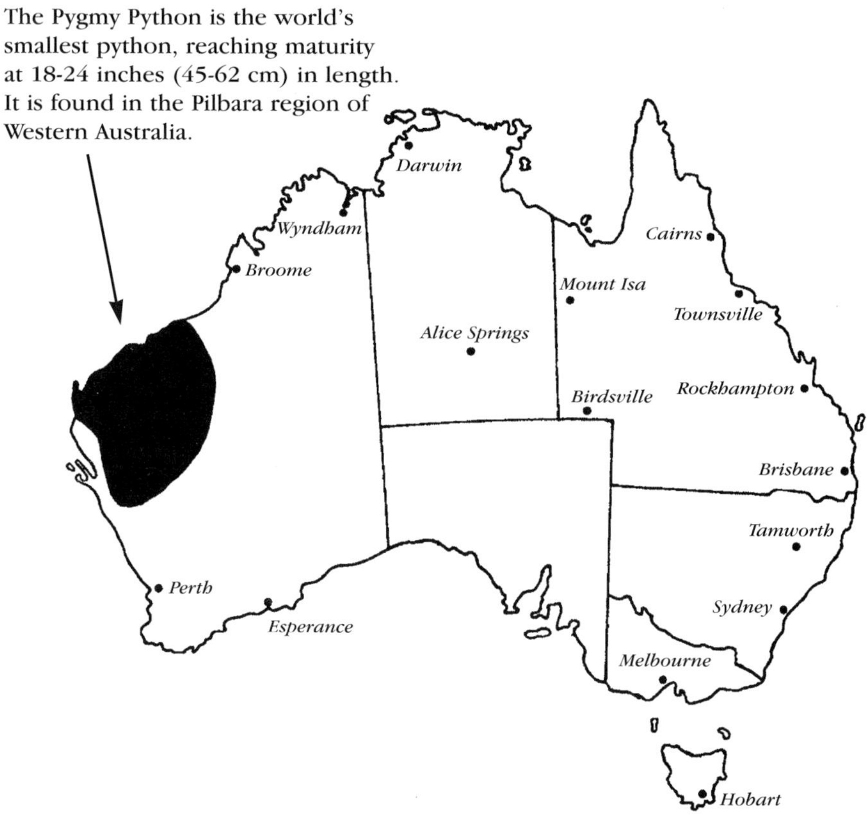

Figure 79. Distribution of the Pygmy Python.

Hoser (1992; 1995) describes night-driving the roads slightly north of Whim Creek, Western Australia to locate Desert Death Adders (*Acanthophis pyrrhus*). In addition to two DOR (dead on road) Western Stimson's Pythons, an adult male Pygmy Python was encountered. The Pygmy Python measured 21.5 inches (55 cm) in length and had about a dozen small ticks on its body. Hoser (1993; 1995) describes locating a resting male Pygmy Python from a small, exposed spinifex clump about 4 miles (6 km) west of Shay Gap, Western Australia.

Clutch size in the Pygmy Python ranges from 2-5 eggs, which are long and ovular (Shine, 1991). Hatching occurs after approximately 60 days of incubation. The neonates are 8 inches (20 cm) in length and well patterned. With age, their pattern fades to the characteristic reddish coloration.

Western Stimson's Python

Antaresia stimsoni stimsoni (Smith, 1985)

The Western Stimson's Python is another small Australian python which reaches adult size at about 2-3 feet (0.6-1.0 m) in length (Figs. 9, 33, 35). This species has an irregular pattern dominated by round blotches or bands, and specimens from different portions of their range are extremely variable in color (Fig. 80). Perhaps due to cooler weather conditions, specimens from the southwestern parts of the range are darker than from other localities. A pale morph with a light grey background and faded dorsal pattern occurs on Dorre Island, Western Australia.[20]

This species is a nocturnal hunter which spends the day sheltered in termite mounds, rock formations, or beneath spinifex bushes (Figs. 19, 22, 26, 34, 36, 47, 64).[20] Interestingly, adult specimens have been found in termite mounds with Pygmy Pythons, although they are not known to hybridize (Hoser, 1993). The Western Stimson's Python is an opportunistic feeder that forages for small mammals, lizards, including geckos of the genus *Gehyra*[19] (Fig. 6).

Figure 80. Distribution of the Western Stimson's Python.

Clutch size in this species is small, generally between 6-10 eggs. Hatching occurs at approximately 60 days. The neonates are about 10 inches (25 cm) in total length and feed on a variety of small lizards.

Eastern Stimson's Python

Antaresia stimsoni orientalis (Smith, 1985)

The Eastern Stimson's Python is found throughout portions of eastern Western Australia, the Northern Territory, western Queensland, and northern New South Wales (Fig. 81). Highly variable in coloration, its dorsal color normally consists of bold, reddish-brown, smooth-edged transverse bars or blotches (Smith, 1985) (Fig. 37). The Eastern Stimson's Python appears to be the second largest species of the *A. childreni* group; specimens 5 feet (1.5 m) in length are known.[11] The average length, however, is 30-36 inches (0.76-1.0 m). This form lives primarily in large rock formations, although specimens have been found in termite mounds, rodent burrows, and under spinifex bushes (Figs. 38, 58). Diet includes lizards, including several species of geckos and skinks, and small mammals (Fig. 6).[10] A specimen encountered crossing a road south of Alice Springs, Northern Territory regurgitated a Rough Knob-Tailed Gecko (*Nephrurus asper*) when disturbed.[15] Like many pythons, this species is observed prowling after warm thunderstorms. In early December, 1994, eight specimens were observed just west of Alice Springs, drinking from small pools of rainwater following a thunderstorm.[7]

Clutches, which vary from 8-15 eggs, are oviposited at the beginning of the wet season. The neonates are about 10 inches (25 cm) in length. In captivity, male specimens show reproductive behavior for 5-7 months, which is longer than in most Australian pythons. Apparently, some mature males only feed for three to four months of the year.[10]

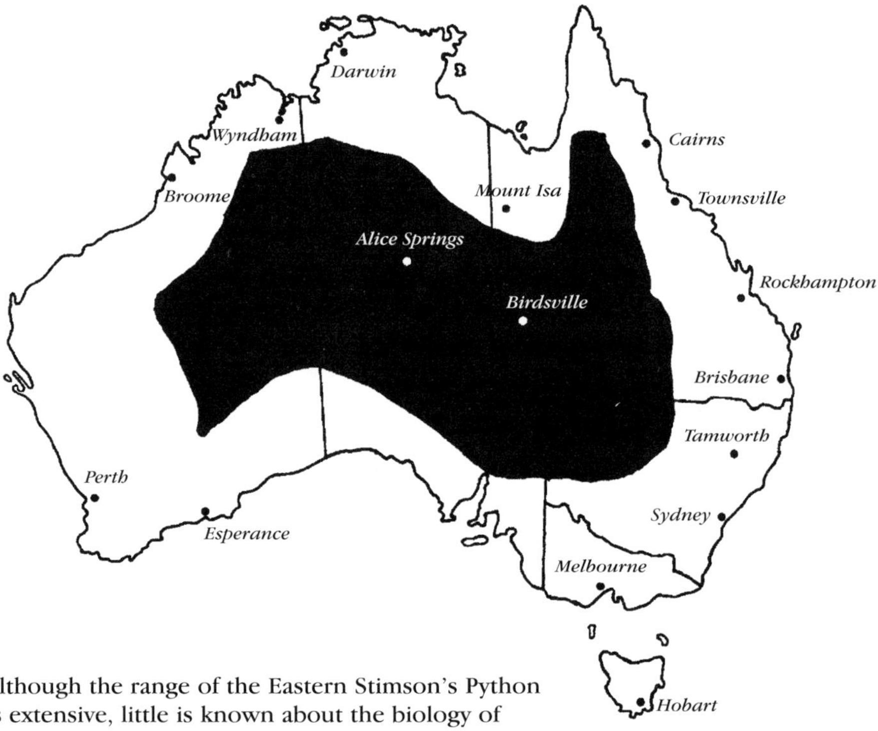

Although the range of the Eastern Stimson's Python is extensive, little is known about the biology of this species.

Figure 81. Distribution of the Eastern Stimson's Python.

Interpreting Relationships in the *Antaresia childreni* Species Group

Prior to the mid-1980s, all forms of the Children's Python were treated as a single taxon, *Liasis childreni*. The common names Children's Python or Banded Rock Python (Kinghorn, 1964) encompassed the group. The Pygmy Python, then called the Western Children's Python, was considered a subspecies of the nominate form and known as *Liasis childreni perthensis*.

Laurie Smith, of the Western Australian Museum, partitioned the populations of *Liasis childreni* into three species, viz., Children's Python (*L. childreni*), Spotted Python (*L. maculosus*), actually first proposed by Peters (1873), and Stimson's Python (*L. stimsoni*), the later taxon with two subspecies, *stimsoni* and *orientalis*. These taxa, along with the Pygmy Python (*A. perthensis*) were indicated to comprise the *L. childreni* group (Smith, 1985).

Smith's (1985) revision and proposed recognition of the discussed species has been largely accepted by subsequent workers, but the description of *L. stimsoni orientalis* is equivocal. Kluge (1993) partitioned *Liasis* into two genera: *Antaresia*, which includes the five members of the *childreni* group, and *Liasis*, which includes the three members of the *olivacea* group.

In this book, I have followed the proposed changes of both Smith (1985) and Kluge (1993). Also, I have accepted the arguments provided by Barker & Barker (1994b) in applying the common name of Northern Brown Python for *A. childreni*, as opposed to the traditional, but problematic common name of "Children's Python." As indicated by Barker & Barker (1994b), the common name Children's Python poses problems due to the amount of data lost from specimens incorrectly identified. For those not familiar with Smith (1985), the common name "Children's Python" can still refer to the entire group or to *A. maculosus*, the most common of the taxa in collections worldwide (Kend, 1992). Historically, the name Children's Python has rarely been used with the species it actually refers to, *A. childreni*. The acceptance of Northern Brown Python as the common name for *A. childreni* will further help to resolve the confusion surrounding this group.

I have chosen to retain the use of the taxon *A. stimsoni orientalis*, the Eastern Stimson's Python (*see* section on **Python Systematics**). Perhaps the most controversial element of Smith's (1985) work is that the subspecies were not clearly differentiated and that all 223 specimens examined were recorded as paratypes (Barker & Barker, 1994b). Ehmann (1992) mentions populations of Stimson's Pythons from the interior regions of New South Wales and South Australia, which he describes as having "...a very robust body with a softer reddish and less contrasting colour with finer patterning." He refers to these populations as the "southern, undescribed race."

There are notable morphological and behavioral differences between the western and eastern populations of Stimson's Python. Differences exist in scalation and coloration, and also in the lateral strip, a pattern character (Figs. 9, 33, 35 vs. Fig. 37). The lateral stripe is a pale dorsal marking formed by an absence of pattern on the neck and anterior portion of the body on some Northern Brown Pythons and most Stimson's Pythons. However, although the lateral stripe is distinct in some Stimson's Pythons, it is not the primary characteristic which differentiates *stimsoni* from *childreni*. The primary character differentiating *stimsoni* is a strongly pronounced pattern throughout life. In contrast, the juvenile pattern of a Northern Brown Python is bold, but fades dramatically with maturity (Barker & Barker, 1994b).

In the Northern Brown Python the lateral stripe is only distinctive in hatchling specimens (Figs. 13, 27). In the Stimson's Python a lateral stripe is typically displayed, but the stripe has a tendency to vary in size and prominence. Specimens of the Western Stimson's Python usually have thick lateral stripes, almost giving the appearance that they were painted on the specimen. However, many Eastern Stimson's Pythons from western Queensland, northern New South Wales, and sometimes South Australia, have less striping. In both taxa, however, the lateral stripe may be apparent in the anterior half of the python (Barker & Barker, 1995a).

Animals from populations of Western and Eastern Stimson's Pythons also behave differently. While most Eastern Stimson's Pythons are calm and docile, many Western Stimson's Pythons tend to be defensive and readily strike when disturbed. Australian herpetologists report that western specimens are more inclined to bite.[22]

More confusing in this group is the occurrence of hybridization areas, which are not well documented or understood. Four such zones are known, and areas where unusual specimens have been reported are indicated (Fig. 82).

(1) An intergradation area between Western Stimson's Python and Eastern Stimson's Python exists throughout central Western Australia along the interface of their ranges (Smith, 1985). (2) A hybrid zone between Northern Brown Python and Western Stimson's Python occurs south of the southern Kimberley of Western Australia.[20] In the Napier Ranges, two particularly distinctive specimens, which may not be hybrids, have been observed by Weigel (*pers. comm.*), who felt that they were distinctive enough to label them an undescribed form of small python. (3) A hybrid zone between Northern Brown Python and Eastern Stimson's Python occurs from the southeastern Kimberley, Western Australia through the Northern Territory's Tanami Desert, ending in the Barkly Tableland and western Queensland (Smith, 1985). This is the most commonly encountered hybrid of the *A. childreni* species group (Figs. 39, 40).[10] (4) A hybrid zone between Eastern Stimson's Python and the Spotted Python occurs in central Queensland, from the base of the Cape York Peninsula south towards Rockhampton. Hybrids from central Queensland have morphological characteristics present in both the Eastern Stimson's Python and the Spotted Python (Hoser, 1993).[2,10]

Future work that includes detailed morphological and molecular analyses is needed to resolve the taxonomic issues that surround this group of pythons.

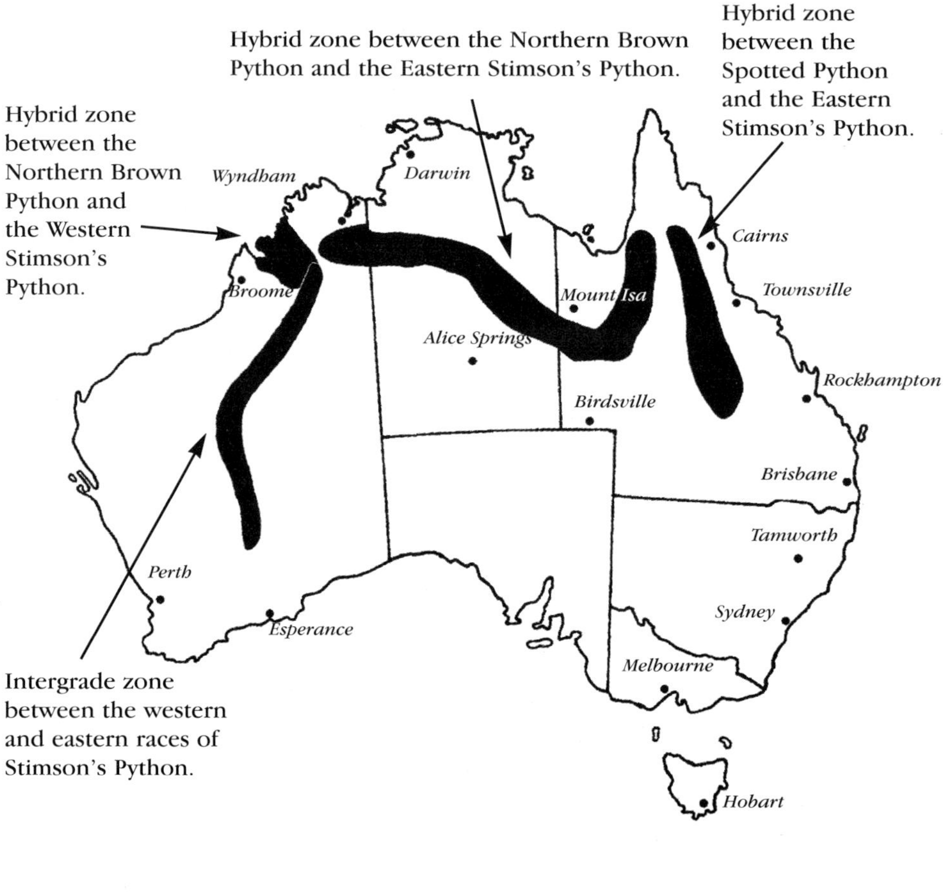

Figure 82. Areas of intergradation and hybridization in the *Antaresia childreni* Species Group.

Genus *Leiopython*
Hubrecht, 1879

The genus *Leiopython* is monotypic; its sole member is the White-Lipped Python, *L. albertisii*. This species is characterized by having a single pair of prefrontal scales. The White-Lipped Python inhabits offshore islands of the Torres Strait, New Guinea, Irian Jaya, and Mussau Island of the Bismarck Archipelago.

White-Lipped Python
Leiopython albertisii (Peters & Doria, 1878)

The White-Lipped Python inhabits the northern islands of Dauan and Saibai in the Torres Strait, which are Australian territory (Figs. 2, 3, 83). There also are unsubstantiated records that it occurs on Badu, Boigu, and Thursday Island (Barker & Barker, 1994b).[2] Moreover, Ehmann (1992) states that this species may be present along the Pascoe River at the northern tip of the Cape York Peninsula, based on sight observations. The White-Lipped Python, however, may be easily confused with the Water Python.

McDowell (1975) reports that populations from the Torres Strait and southern Papua New Guinea have a dorsum color of grey and copper-brown with a yellow venter. Cogger (1992) indicates that Australian specimens are overall grey-brown, and that the young sometimes have a dark head. Further, he reported the average length for specimens as 6.5 feet 2(2 m), with a maximum length of 9.75 feet (3 m).

This python inhabits forested areas with dense underbrush and coastal areas of scattered mangrove and shrub. Loveridge (1948) notes hair of the bandicoot (*Isodon* spp.), a small marsupial, in the stomach of an "Australian" specimen; McDowell (1975), Shine & Slip (1990), and Barker & Barker (1994b) report that it feeds on rats.

Little information is available on the natural history of the White-Lipped Python from the Torres Strait, and only a handful of specimens are represented in museum collections. I was unable to locate a suitable

Figure 83. Distribution of the White-Lipped Python in Australia.

photograph of an Australian specimen in its natural habitat, so it is not illustrated here (see Ross & Marzec, 1990, Cogger, 1992, and Barker & Barker, 1994b for photos of this species from other areas of its range).

Genus *Liasis*

Gray, 1842

The genus *Liasis* occurs throughout most of Australia, Indonesia, New Guinea, and surrounding islands. It differs from other python genera by the presence of regular shield-like head scales and by having one (occasionally two) loreal scales.

Members of the genus with an extralimital range include the Water Python (*L. fuscus*), while non-Australian species include the Indonesian Water Python (*L. m. mackloti*) and the Savu Island Python (*L. m. savuensis*).

Key to the Australian Species of *Liasis*

1.	More than 350 ventral scales	2
	271-286 ventral scales	*fuscus*
2.	355-377 ventral scales	*olivaceus olivaceus*
	374-411 ventral scales	*olivaceus barroni*

Water Python

Liasis fuscus (Peters, 1873)

The Water Python is a large, semi-aquatic species found throughout northern Western Australia, the Northern Territory, and Queensland (Fig. 84). Attaining a length of up to 9 feet (3.0 m), this python is a major predator in northern Australia's floodplains. For a patternless snake, the Water Python is variable in color, which ranges from dark brown, to golden brown and greyish-silver (Fig. 41). Some specimens display three distinctive ventral colors, with white under the head, bright yellow or orange ventrally to the cloaca, and brown underneath the tail.

Throughout northern Australia, many populations of the Water Python are fully aquatic as a result of their microhabitat flooding during the wet season and months thereafter.

Figure 84. Distribution of the Water Python.

The Olive Python is sometimes mistaken for the Water Python, but the two are easily differentiated: the Water Python has yellow while the Olive Python has white ventral scales.

The Water Python is generally nocturnal, and frequently shelters in hollow logs, wood piles, or holes in the ground during the day (Figs. 7, 42, 54). At dusk these pythons emerge to prowl for food.[9] This species has a varied diet, and is suspected of eating flying foxes (*Pteropus sp.*), water birds, young Magpie Geese (*Anseranus semipalmata*) and their eggs (Shine, 1991) (Fig. 6). The Floodplain Rat (*Rattus colletti*) constitutes the bulk of its diet (Shine, 1992). Cogger & Lindner (1974) reported that an adult Water Python regurgitated two juvenile and one adult Bandicoots (*Isodon macrourus*). Some authors, including Stafford (1986) and Gow (1989), have indicated that the Water Python's diet includes young Salt-Water Crocodiles (*Crocodylus porosus*). Shine (*pers. comm.*) has studied this species in the field for over a decade and has never observed predation by Water Pythons on Saltwater Crocodiles (*Crocodylus porosus*). Shine (1992) states that young Water Pythons also prey on small snakes, particularly the Slate-Grey Snake (*Stegonotus cucullatus*).

The ecology of the Water Python is under study by Dr. Richard Shine, Dr. Thomas Madsen, and colleagues (Madsen & Shine, 1996a; Madsen & Shine 1996b). Through the use of radio transmitters implanted in the body cavity, these researchers have discovered interesting ecological aspects of the Water Python at Fogg Dam in the Northern Territory. First, Water Pythons search for their prey and are not ambush predators. Second, they are able to adjust to a dramatic change in conditions and available prey caused by seasonal shifts in water availability. During the dry season, the pythons spend their days coiled under the reed beds of the billabong, emerging at night to hunt for rats in the dry floodplain. The pythons are forced to become fully aquatic with the coming of the wet season. Much of its habitat, however, consists of creeks, swamps, and lagoons that are not flooded seasonally.

Following the breeding season, female Water Pythons will oviposit 12-22 eggs, generally choosing paperbark tree roots or floodplain burrows as nesting sites (Barker & Barker, 1994b).[21] Barker & Barker (1994b) report on 15 clutches ranging from 6-23 eggs. Neonates average 18 inches (35-45 cm) in length.

Olive Python

Liasis olivaceus olivaceus (Gray, 1842)

The Olive Python is found throughout northern Western Australia, the Northern Territory, and Queensland (Fig. 85). Relatively uniform in coloration, the only visual distinction between specimens is in the contrast of the olive-brownish colored dorsal scales (Fig. 43). This is one of Australia's larger pythons, with an average length of 8-10 feet (2.46-3.07 m). Specimens of 15 feet (4.6 m) in length, however, are occasionally reported.[22]

This python inhabits rocky hillsides, particularly if the outcrops are near water (Figs. 7, 44, 52, 54, 71). In open grassland or forest, they frequently shelter among hollow fallen trees and large termitaria. They appear to be active at dusk and often are encountered crossing roads in the Northern Territory. Prey includes both mammals and reptiles. Large adults will eat Agile (*Macropusa agillis*) and Rock Wallabies (*Petrogale* spp.), which are about the size of a large domestic cat. Specimens from the Kimberleys, Western Australia, often show extensive scarring. This could due to their choice of prey, as Mertens' Monitors (*Varanus mertensi*), Freshwater Crocodiles (*Crocodylus johnstoni*), Australian Native Cats (*Dasyurus hallucatus*), Rat Kangaroos (*Bettongia lesueur*), and rock wallabies (*Pterogale* spp.) are found along the same intermittent pools in the King Leopold Ranges.[1]

The Olive Python may be ophiophagous. In April 1982, a large (8 feet; 2.46 m) specimen was observed constricting a 30 inch (0.76 m) Northern Brown Python in the Napier Ranges, Kimberleys, Western Australia. The Olive Python, however, appeared to have difficulty killing the extensively lacerated Northern Brown Python.[21] In captivity, a 10 foot (3.0 m) specimen devoured its cagemate, a 6 foot (1.8 m) Central Carpet Python.[10] Still, ophiophagy is not confirmed, as many captive snakes become highly excited during feeding. This behavior, typical of many pythons, may explain why an Olive Python swallowed a tube sock containing insecticide used to kill snake mites, which was placed overnight in its enclosure.[22] In captivity, Olive Pythons appear to reach sexual maturity in 3-4 years, although the majority of captive breedings have occurred with specimens 8-12 years of age.[22] Following the mating season, up to 35 eggs are laid. Barker & Barker (1994b) report on 6 clutches ranging from 14-31 eggs. The neonates hatch 80-100 days later and are 18-24 inches (46-61 cm) in length.

Pilbara Olive Python

Liasis olivaceus barroni (Smith, 1981b)

The Pilbara Olive Python is found throughout the Pilbara region of Western Australia (Fig. 85). It can attain a length of over 20 feet (6.15 m) and is Australia's heaviest snake. Most mature specimens encountered in the field are about 10 feet (3 m) in length.[1] Whitlock (1923) collected a specimen measuring 17.8 feet (5.50 m) during a trip to the Fortescue River and Hamersley Ranges. Pilbara Olive Pythons are a dusty olive-brown, copper, or red in coloration (Figs. 45, 46). Specimens from the Chichester Ranges are dark in coloration and have large, blunt heads.[1] Generally, the Pilbara Olive Python can be distinguished from the Olive Python on the basis of color. Some Pilbara Olive Pythons, however, cannot be accurately distinguished from the nominate form on the basis of color alone. In such cases, the number of ventral scales can be used to differentiate the races (*see* **Key to the Australian Species of *Liasis***). Further, the Pilbara Olive Python may display a white stripe stretching between the eye and the posterior head.[12] Specimens from the Hamersley Ranges observed at night on rock outcrops near permanent waterpools have exhibited a red eyeshine.[12]

Adult Pilbara Olive Pythons inhabit gorges and rocky outcrops surrounding permanent water, where they move to crevices when disturbed (Figs. 22, 32, 36, 47). In this arid region, permanent waterpools are rare and are the focal points for local wildlife. This python has been observed to lie near or in the water and feed on water birds and various mammals that come to drink.

The Olive Python inhabits the rocky hillsides, watercourses, and forested savannas of northern Australia. It is Australia's fourth longest python. Specimens display an olive-brown coloration.

Olive Python
(Liasis olivaceus oliaceus)

Pilbara Olive Python
(Liasis olivaceus barroni)

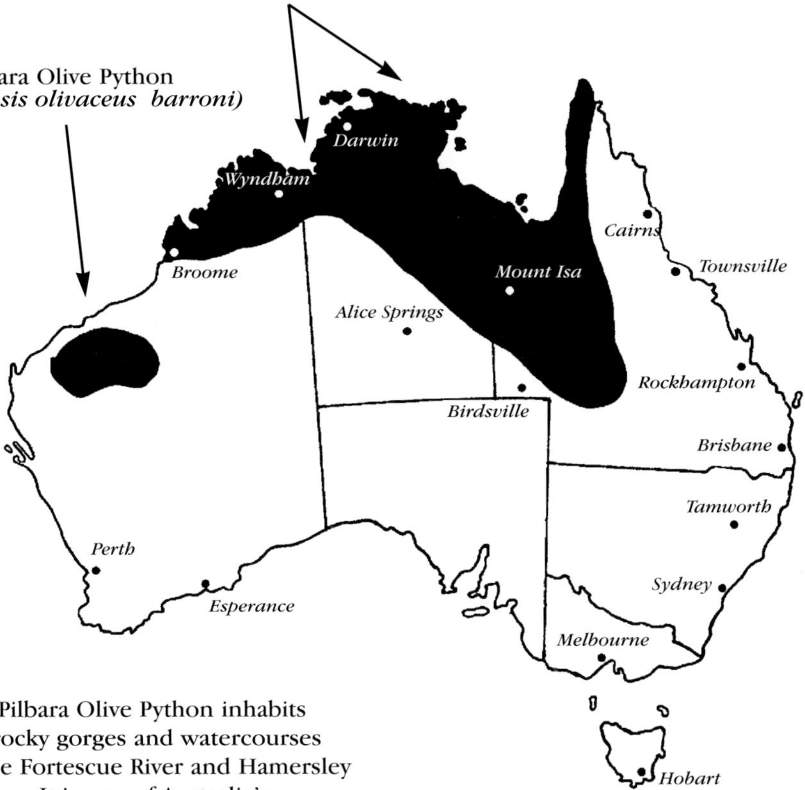

The Pilbara Olive Python inhabits the rocky gorges and watercourses of the Fortescue River and Hamersley Ranges. It is one of Australia's largest snakes. Specimens generally display a dusty red coloration.

Figure 85. Distribution of the Olive Python.

The population status of this large python needs further investigation. In captivity, I am only aware of a single male specimen. The snake measures about 11 feet (3.38 m) and weighs approximately 33 lbs (15 kg). Eighteen months earlier, soon after its capture, it measured slightly over 6 feet (1.85 m). The specimen was severely scarred, perhaps from intense male-male combat, but this seemed unlikely based on the size of the animal (e.g., immature). While sexing the specimen, however, a seminal plug was noted. As stated by Mengden (1980), a seminal plug indicates reproductive activity.

Interpreting Relationships in the *Liasis olivaceus* Species Group

The *Liasis olivaceus* species group is comprised of *L. fuscus, L. mackloti,* and *L. olivaceus*. There has been some disagreement, however, over the taxonomic status of *L. fuscus* and *L. mackloti*. The Indonesian Water Python (*L. m. mackloti*) was described by Dumeril & Bibron (1844) from Timor, Indonesia. This taxon ranges throughout Samao, Sawoe, Timor, the Wetar Islands, and the Lesser Sundas (Kluge, 1993) (Fig. 2). The Water Python (*L. fuscus*) was described by Peters (1873) from a specimen collected at Port Clinton, Queensland, and ranges throughout northern Australia and southern New Guinea (Kluge, 1993) (Figs. 2, 84).

Both species were considered distinct until McDowell (1975) stated that a taxonomic key differentiating *L. fuscus* and *L. mackloti* in Boulenger (1893) was not useful. McDowell examined ten specimens, four *L. fuscus* from southern New Guinea, four *L. fuscus* from Australia, and two *L. mackloti* from Wetar Island. He acknowledged, however, that there were detectable differences between Australian and southern New Guinea *L. fuscus* from Lesser Sundas *L. mackloti*. Kluge (1993) recognized only *L. mackloti*, stating that only one historical entity was present.

I agree with the opinion of Barker & Barker (1994b) that *L. fuscus* and *L. m. mackloti* are distinct species. In appearance, behavior, and distribution, there are obvious differences between the two forms. For example, *L. fuscus* is a patternless snake and displays a range of dorsal colors: golden brown, dark brown, black, and grey-silver. In contrast, *Liasis m. mackloti* is mottled or speckled, and its dorsal ground color is olive-grey, pale grey, grey, dark brown, or black. The behavior of these taxa also is distinctive. Both neonate and adult *fuscus* have been reported to range from being gentle to irritable. Barker & Barker (1994b) reported hatchling *L. fuscus* to be frequent biters. Based on a number of observations (L. Porras, *pers. comm.*), a few specimens of young *L. fuscus* from southern New Guinea were hesitant to strike defensively or bite. This behavior was consistent with that of juveniles from several clutches of *L. fuscus* from Australian parental stock, but differed from that of *L. mackloti*, where neonates from several clutches were extremely nervous and defensive. I have found adult captive *L. fuscus* to be unpredictable in biting behavior, whereas imported adult *L. mackloti* are among the most gentle of any python. In conclusion, *L. fuscus* and *L. mackloti* are poorly understood taxonomically and in need of further investigation. The same may be said of the relationship between *L. m. mackloti* and the extralimital *L. m. savuensis*.

Genus *Morelia*
Gray, 1842

The genus *Morelia* is found thoughout much of Australia, and parts of Indonesia and New Guinea. The group is differentiated from other genera by the presence of small, irregular scales on the head, six or more labial pits on the infralabials, a prehensile tail, and the presence of teeth on the premaxilla.

Members of *Morelia* with an extralimital range include the Scrub Python (*M. amethistina*), Green Python (*M. viridis*), and New Guinea Carpet Python (*M. spilota variegata*). The only non-Australian member is the Boelen's Python (*M. boeleni*) from New Guinea.

Key to the Australian Species of *Morelia*

1.	dorsal scales smooth	2
	dorsal scales rough, or keeled	*carinata*

2.	Central head scales large and regular	3
	Central head scales small and irregular	4
3.	Parietal shields not fragmented	*amethistina*
	Parietal shields fragmented	*oenpelliensis*
4.	Fewer than 90 subcaudal scales	*spilota*
	90-110 subcaudal scales	*viridis*

Note: At this time there are no morphological characters that will reliably differentiate any of the proposed subspecies of *Morelia spilota* (*see* **Interpreting Relationships in the *M. spilota* Group**).

Scrub Python

Morelia amethistina (Schneider, 1801)

The Scrub Python inhabits the coastal rainforests of northern Queensland (Fig. 86). This is unquestionably Australia's longest snake. Most mature specimens seen in the field are 10-15 feet (3.0-4.5 m) in length, (Figs. 14, 48, 49), and a specimen measuring 28 feet (8.53 m) was recorded in 1948 (Worrell, 1963).

These pythons are mainly found on the rainforest floor and usually in the vicinity of water (Fig. 30, 50, 62, 73). During the daylight hours of summer, Scrub Pythons will coil tightly among the forest floor's fallen branches, leaf litter, or dense vegetation. In winter, groups of Scrub Pythons are known to emerge from the rainforest to bask in the morning sun in exposed clearings or on top of large rock formations. During this time, Scrub Pythons are easily located in the field, particularly in boulder-strewn gorges. While traveling by cable car from the station to the top of the gorge, workers from Tully Power Station in Queensland commonly see large Scrub Pythons basking in the trees and boulders.

The Scrub Python's natural diet consists mainly of marsupials and birds, but with the encroachment of humans into their habitat, also includes small domestic dogs, cats, and poultry (Figs. 6, 14). Adult specimens have been noted to devour tree kangaroos (*Dendrolagus lumholtzi* and *D. bennettianus*).[17] The diet of neonates is mainly small

The Scrub Python inhabits the coastal rainforest of northern Queensland. A specimen measuring 28 feet (8.53 m) was recorded in 1948, making this Australia's largest snake.

Figure 86. Distribution of the Scrub Python in Australia.

mammals and birds. Young specimens occasionally are encountered in bird's nests, tightly coiled with a distended midbody.[14]

In nature, reproduction in Scrub Pythons is highly seasonal. They normally mate from late July to early September, at which time males often are found with large wounds resulting from intense combat, probably over females.[22] After mating, females generally oviposit clutches of 8-16 eggs. Barker & Barker (1994b) report five clutches ranging from 5-21 eggs. The neonates are about 24 inches (61 cm) in length. Young Scrub Pythons undergo color ontogeny, turning from a uniform copper-red to hues of gold, purple, brown, yellow, and black as adults.

Rough-Scaled Python

Morelia carinata (Smith, 1981a)

The Rough-Scaled Python is one of Australia's least studied pythons: only a few individuals have been found, and neonates remain unreported. Named by Smith (1981a), the first Rough-Scaled Python, an adult male, was found on 14 January 1973, at Mitchell River Falls, Western Australia (Fig. 52). This specimen, slightly under 6.5 feet (2.0 m) in length, was collected near Admiralty Gulf close to a fresh water creek.[20] In 1987, a second specimen was collected by Laurie Smith and Ron Johnstone during a museum survey of Prince Frederick Harbour, approximately 50 miles west of Mitchell River Falls. The specimen, a young male which measured about 4 feet (1.2 m), was collected from a small patch of monsoon rainforest at the mouth of the Hunter River in Prince Frederick Harbour (Fig. 51). Unfortunately, at the time of collection, both specimens were mistaken for Northern Carpet Pythons and subsequently sacrificed for museum collections (Weigel & Russell, 1993). The first specimen was shot out of the tree where it was discovered.[21]

The third Rough-Scaled Python was discovered at Prince Frederick Harbor, at the source of the Hunter River, by John Weigel and Trent Russell of the Australian Reptile Park in Gosford, New South Wales on 12 June 1993. This specimen, a young male, was about 30 inches (80 cm) in length. As reported by Weigel & Russell (1993), it

The elusive Rough-Scaled Python is known from only a few specimens. Its range is restricted to the northwest corner of the remote Kimberleys of Western Australia, the smallest range of any Australian python.

Figure 87. Distribution of the Rough-Scaled Python.

was found in a canyon of scattered monsoon rainforest with assorted fruiting trees and boulders enclosing a slow flowing creek.

Nearly a year later, in mid-May of 1994, John Weigel and Trent Russell journeyed back to Prince Frederick Harbour in search of additional specimens. Again, they were successful, finding an individual about 4.5 feet (1.38 m) in length at 2100 h, about 3 h after sunset. Barker & Barker (1994b) reported that this specimen was spotted 5 feet (1.54 m) off the ground while in an ambush position in a fruiting Lilli Pine Tree. The habitat was a thicket of deciduous rainforest surrounded by rocky outcroppings, about 325 feet (100 m) away from a moving stream. This snake, sex not determined, resembled the third collected specimen in that its dorsal pattern was fawny red-brown and mottled cream.[21] Barker & Barker (1994b) reported (*pers. comm.* with J. Weigel) that the specimen struck slowly with an open mouth and long teeth exposed. After striking the snake remained fully extended and moved its head sideways for about five seconds before closing its mouth and withdrawing the head. This behavior was previously unreported in the genus *Morelia*.

The Rough-Scaled Python's habitat was thought to center around the many sandstone boulders and outcroppings common to this region, but it is now known to occur in patches of monsoon rainforest, scattered in close proximity to the spectacular rock formations of the northwestern Kimberleys (Fig. 87).[21] Despite the fact that only a few specimens are known, this species should not be considered endangered. Its distribution is in a very remote part of the continent, mainly accessible by boat or helicopter, which certainly limits human activity in the region. As stated by Weigel & Russell (1993), their habitat is "radical and imposing." Considering that only several Northern Carpet Pythons and a handful of other snakes have ever been recorded in this area, it is likely that future field work will show stable populations of the Rough-Scaled Python. Also, since the holotype was found about 50 miles west of Prince Frederick Harbour, where the other three specimens were located, it seems likely that this species might range throughout more of the forested creeks, gorges, and waterways of the northwestern Kimberley.

Oenpelli Python

Morelia oenpelliensis (Gow, 1977)

The Oenpelli Python is found in the Arnhem Land escarpment of the Northern Territory (Fig. 88). Reaching a length of 10-16 feet (3.0-5.0 m), the Oenpelli Python is Australia's third longest snake (Figs. 8, 11, 53). Gow (1989) records a specimen measuring 14.86 feet (4.572 m) in length. This species is extremely slender for its length. Krauss (1992a) reported specimens measuring 10 and 12 feet (3.0-3.8 m) which weighed 7.7 and 13.3 lbs (3.5-6.0 kg), respectively. The ratio of length to weight in the Oenpelli Python is far different from that of other large pythons. For example, an Indian Python (*Python m. molurus*) of similar length would weigh over 30 lbs (13.63 kg).

Inhabiting fissured sandstone escarpments, caves, and both rocky and forested gorges (Figs. 7, 54), the Oenpelli Python feeds on large marsupials and birds.[11] Gow (1989) states that Aborigines report prey as large as the Black Euro (*Macropus bernardus*) is taken. Oenpelli Pythons also eat flying foxes (*Pteropus* spp.) and native pigeons (*Columbigallina* spp.) (Fig. 8). They also frequent fruit producing trees for several months of the year when prey is available.[11]

Peter Krauss of Queensland, Australia witnessed copulation in the Oenpelli Python on several occasions, and subsequently a clutch of nine eggs was laid, two of which hatched. The young pythons measured 32 and 35 inches (83 and 90 cm). It is interesting to note that the neonates did not shed their skin until 66 days after hatching (Krauss, 1992).

This python may produce the largest eggs of any Australian python, although data on egg size from the Pilbara Olive and Rough-Scaled pythons are not available. Mean mass of Oenpelli Python eggs was 185.6 g (Charles & Wilson, 1985). This is nearly twice the mass of Scrub Python eggs, which generally average about 100 g. The large size of Oenpelli Python eggs has led to speculation that further research will demonstrate this taxon to be Australia's longest snake. Large snakes, however, do not necessarily produce the largest eggs and offspring (Gow, 1989).

The Oenpelli Python is restricted to the craggy sandstone escarpments of eastern Kakadu National Park and western Arnhem Land, Northern Territory. It is Australia's third longest snake. It was formally described in 1977.

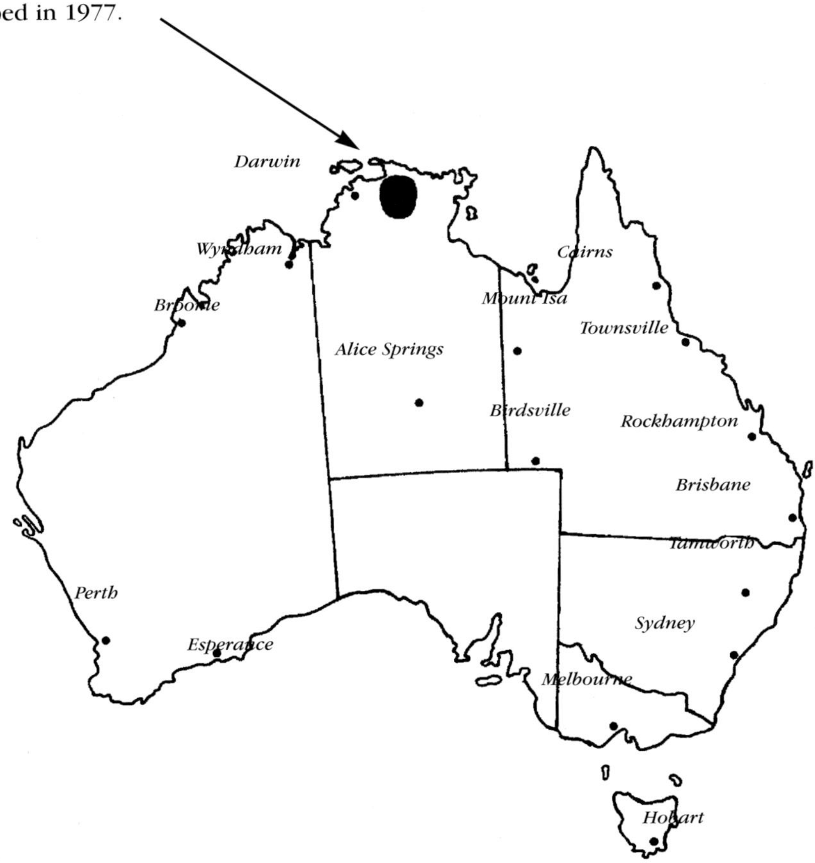

Figure 88. Distribution of the Oenpelli Python.

Diamond Python

Morelia spilota spilota (Lacepede, 1804)

The Diamond Python is found along the southern half of the New South Wales coast and hinterland, and is one Australian python that lives in a temperate zone (Fig. 89). Sometimes exceeding 9 feet (3 m) in length, it is regarded as one of the larger members of the *M. spilota* group. A specimen measuring 12 feet, 1 inch (3.68 m) was recorded in 1971.[22] Its dorsum is olive black in overall coloration, and its scales have white or yellow spots (Figs. 15, 55). Subtle variation in color occurs, though not as much as in other Carpet Pythons. When removed from the wild and kept indoors, the color of the Diamond Python tends to fade. The yellow pattern will usually turn white after a few months.[22]

This python is often encountered in the semi-forested valleys of Sydney's suburbs (Figs. 16, 56). It was in this area that the first radiotelemetry study on pythons was conducted (Slip & Shine, 1988a-f). It was found that during summer months these pythons often travel through cooler valleys in search of food or mates. In winter they normally move to higher elevations, occupying refuges or basking sites that face east, north-west, or west, so as to keep warm. Juveniles tend to shelter under smaller rocks at the top of rocky cliffs. Here they are warmed by sun-baked rocks without the danger of exposing themselves to predators.

Slip & Shine (1988d) found that adult Diamond Pythons generally prey on rats (*Rattus fuscipes* and *R. rattus*), Brushtail Possums (*Trichosurus vulpecula*), and rabbits (*Oryctolagus cuniculus*), and juveniles to prey on Copper-Tailed Skinks (*Ctenotus taeniolatus*), Sugar Gliders (*Petaurus breviceps*), and mice. Slip & Shine (1988d) observed several adult Diamond Pythons feeding in the field. One specimen, was seen busy devouring an adult Brushtail Possum about 19.5 feet (6 m) up a large gum tree. Another specimen was observed lying in the leaf litter at the base of a large pine tree (*Casuarina*), feeding on an adult Ringtail Possum (*Pseudocheirus peregrinus*).

Free-ranging female Diamond Pythons may breed every third year (triennial cycle) (Slip & Shine, 1988f). Normally, they mate from August to October, with oviposition occurring from December through

January. They often build a nest out of leaf litter, generally at the base of a bush or tree. It is suggested that this behavior helps provide more insulation for incubating females (Shine, 1991). Females also have been observed to leave their nest to bask until their body temperature reached 91°F (32.7°C). Through a combination of twitching and basking, females incubating their eggs can elevate their body temperature up to 32°F despite cooler air temperatures (Slip & Shine, 1988f).

Hatching occurs after about 65 days of incubation, usually in February or March. The average clutch size is 20-25 eggs, but can exceed 40. Shine (1988f) reported a clutch of 54 eggs. Barker & Barker (1994b) reported 12 clutches ranging from 12-54 eggs. Neonate Diamond Pythons measure 12-14 inches (30-36 cm) in length. The Diamond Python is considered by some to be one of the most difficult boids to successfully maintain and reproduce in captivity. Barker & Barker (1994b) reported that adults often suffer from numerous chronic health problems, particularly respiratory infections. These ailments, which generally prevent them from living more than eight years in captivity, may result from improper captive conditions.

In my opinion, there are three major elements necessary for the proper maintenance of the Diamond Python in captivity. The first involves the degree of temperature variation specimens are exposed to during general maintenance and breeding trials. In nature, the temperature routinely falls to 40°F (4°C) during the winter. Thus, maintaining specimens at uniform or excessively warm temperatures has been largely unsuccessful. Ross & Marzec (1990) list winter temperatures for their habitat at 65-70°F (18.3-21.1°C) as day-time highs and 40-45°F (4.0-7.2°C) as night-time lows. They also suggest that males should be cooled to 60-65°F (15.6-18.3°C) during the breeding season.

During cool weather, Diamond Pythons often bask to elevate their body temperature. Observations by Slip & Shine (1988a) show that their body temperature is quickly elevated through basking. Captive experiments have demonstrated that they are able to reach body temperatures of 91°F (32.7°C) through basking when ambient temperatures are 65°F (18.3°C) (Ross & Marzec, 1990). Furthermore, basking may have a secondary function: exposure to ultra-violet light

(UV-A, UV-B). Lengthy exposure to ultra-violet light, which is not necessary for most captive snakes, may be necessary for the long-term maintenance of this python (e.g., skeletal development).

The last element involves the lack of multiple adult specimens in a breeding group. A key part of the Diamond Python's reproductive biology are mating aggregations, generally composed of one female and several males. Male Diamond Pythons do not combat. Mating aggregations, however, enable females to be bred numerous times, potentially with numerous partners, which may increase the probability of fertile clutches (Shine, 1991). In captivity and outside Australia, where Diamond Pythons command a steep commercial price, most breeders do not maintain multiple male specimens to allow for such aggregations. This, in part, may account for the high percentage of infertile clutches reported by Ross & Marzec (1990).

In Australia, propagation of the Diamond Python has occurred most frequently in large, outdoor enclosures. Many Australians believe that specimens are best maintained where they can expose themselves to natural sunlight and select from a variety of thermal areas (i.e., thermoregulate).

Central Carpet Python

Morelia spilota bredli (Gow, 1981)

The Central Carpet Python is found throughout the MacDonnell ranges and nearby areas of the southern Northern Territory (Fig. 89). In nature, these pythons are usually bright red-orange (Fig. 57). During their first three months in captivity, however, wild-caught specimens have a tendency to become dark red-brown (Fyfe, 1990). This abrupt change, possibly related to the lack of ultra-violet light when specimens are maintained indoors, should be investigated. The Central Carpet Python reaches 6-8 feet (2-2.45 m) in length; one specimen was measured at 8.5 feet (2.6 m).[10]

The habitat of the Central Carpet Python includes tree-lined rivers and creeks, particularly near rocky hillsides (Fig. 58). Common refugia include hollow logs, rock crevices, and rabbit burrows. Fyfe (1994) reports specimens drinking from puddles after thunderstorms. Like all *M. spilota*, this python exhibits arboreal tendencies. Feeding, mating, and maternal incubation all have been observed in trees. Adults prey on a variety of mammals, including feral cats, rabbits, and birds. Large prey, such as the Rock Wallaby, have been reported (Fyfe, 1990). The diet of neonates and juveniles consists mainly of geckos, particularly members of the genus *Gehyra.*

Fyfe (1994) reported that male Central Carpet Pythons often bite each other during combat, and, by looking for healed wounds, it is often possible to determine the sex of wild individuals. The Central Carpet Python generally oviposits 16-20 eggs. Seventy hatched python eggshells were found in a tree hollow, evidence that more than one female may use the same incubation site or that the site has been used for more than a single year (Fyfe, 1990). Barker & Barker (1994b) reported six clutches which ranged from 13-47 eggs. Hatching occurs at about 60 days. The neonates are approximately 16 inches (39.83 cm) in length, and their pattern resembles that of the adults, except that they are grey, and not red, in color.

Jungle Carpet Python

Morelia spilota cheynei (Wells & Wellington, 1984)

The Jungle Carpet Python is found throughout the coastal regions of northern Queensland (Fig. 89). Barker & Barker (1994b) state that this python is found along the Barron, Mulgrave, Johnstone, and Russell rivers and their tributaries, all which drain the Atherton Tablelands, as well as along the Daintree, Tully, and Herbert River drainages. South of the reported range, Jungle Carpet Pythons occur in the forested hills west of Mackay, Queensland.[13] Montgomery (1992) notes "jungle-like" specimens, displaying dark chocolate and gold patterning, from the rainforests of the Lockerbie Scrub and Jardine River area on the northern tip of the Cape York Peninsula. In the northern part of the range, specimens banded in white or grey and black are found, and often are referred to as "black and white jungle carpet pythons." Barker & Barker (1994b) stated that these pythons are found near Julatten, and throughout the watercourses inland from Port Douglas.

North of Julatten, unusual specimens have been reported along the closed canopy rainforests and waterways of Cape Tribulation National Park, on the southern Cape York Peninsula near the Daintree River (Fig. 61).[17] At Noah Creek, Mile Creek, and Thompson Creek, specimens with contrasting black and white stripes, and in some cases bands of black and ivory white are encountered. At Alexandra Bay specimens are banded in black and yellow, and strongly resemble young New Guinea Carpet Pythons from the border of southern Irian Jaya and Papua New Guinea. Field observations indicate that these pythons average about 40 inches (1 m) in length, and no recorded specimens have exceeded 6.5 feet (2 m). Lyall Naylor, a park ranger at Cape Tribulation, believes that the tall mountain ridges above these waterways may serve as barriers, thus isolating these unusual pythons to pockets of lowland rainforest. In Cedar Bay, 60 km N of Cape Tribulation, there has been only one sight record of a Carpet Python in 25 years. Possibly isolated from Cape Tribulation due to surrounding dry *Eucalyptus* plain forest, it is unknown why specimens have not been found at Cedar Bay, which has a habitat similar to Cape Tribulation.

The typical Jungle Carpet Python is often beautifully marked, with bright, alternating bands of yellow or gold on a black background (Figs. 59, 60). In nature, most specimens are small, averaging 4-5 feet (1.2-1.5 m) in length.[22] Barker & Barker (1994b) state that Jungle Carpet Pythons along the Johnstone River drainage usually are under 4.5 feet (1.4 m) in length, yet specimens from near Atherton can attain a length of over 6 feet (1.9 m); west of Atherton, they state that "jungle-like" specimens may reach a length of over 8 feet (2.5 m).

Like all *M. spilota*, the Jungle Carpet Python has strong arboreal tendencies and it is commonly found on forested hillsides and in patches of rainforest (Fig. 62). Wide deforestation and agricultural development have occurred over much of its range. Parts of the Atherton Tablelands and surrounding areas have been cleared to produce crops of maize and peanuts, as well as supporting timber, pigs, and dairy farms. In some areas, these activities may have restricted this python to small patches of remaining rainforest. Nonetheless, this python is not under the threat of extinction as stated by Wells & Wellington (1984). In Julatten and nearby areas, these snakes are frequently observed basking in windrows, or clearings of chopped-down trees that serve as borders between farmland and upland rainforest. Although most windrows are eventually burned to make way for farmland, they become important nesting sites for Mosaic Tailed Rats (*Melomys* spp.), and serve to attract the pythons.[17]

In nature, this species is known to lay clutches of 10-12 eggs.[22] Barker & Barker (1994b) reported 30 clutches of 9-28 eggs. Neonates are usually dull black and white upon hatching (Montgomery, 1992) and begin to develop yellow coloration when they are from 6-12 months of age.

Southwestern Carpet Python

Morelia spilota imbricata (Smith, 1981a)

The Southwestern Carpet Python is native to the southwest corner of Western Australia, but is found also on several islands off the coast of South Australia along the Great Australian Bight (Figs. 3, 89). Adult specimens are usually between 4-7 feet (1.2 - 2.0 m) in length. Most are green-black in color, but can vary slightly in dorsal color and pattern (Fig. 63). Rocky outcroppings, particularly near tree-lined rivers and creeks (Fig. 64), are common refugia. This snake is a good climber, and often conceals itself in hollow logs and limbs (Bush, 1981). From what little is known, it is assumed that these pythons prey on small mammals and birds. Farmers often release them near grain sheds in an attempt to control rodents.[3]

This python also is found on offshore islands; there they undergo seasonal shifts in feeding, comparable to those of the large Black Tiger Snakes (*Notechis ater serventi*) of Mount Chappell Island in Bass Strait (Shine, 1991). Here both species prey primarily on Muttonbird (*Puffinus tenuirostris*) chicks, but because of the birds' nesting habits, the snakes have five weeks (from early January to February) to feed on the birds before they are gone. Afterwards, they feed sporadically on small lizards and rodents, particularly the Bush Rat (*Rattus fuscipes*).[18]

Concern has been expressed about the conservation of these insular pythons. Schwaner et al. (1988) state that the population on Thistle Island, discovered in 1802, is probably extirpated, and it is no longer present on North Wallabi Island (Cogger et al., 1993). On St. Francis Island, the population is estimated at 500-1000 specimens, and it is possible that fishermen use these pythons as bait in craypots (Schwaner et. al., 1988). Cogger et al. (1993) listed this taxon as threatened and initiated a detailed recovery plan under the Action Plan for Australian Reptiles.

The Southwestern Carpet Python has been successfully bred in captivity. Barker & Barker (1994b) report two clutches of 17 and 20 eggs, with neonates averaging approximately 16.4 inches (42 cm) in length. Bush (1988) noted that a long-term captive female, which measured slightly under 5 feet (1.5 m), oviposited 14 eggs. The eggs, incubated in damp vermiculite, were 51-55 mm in length and 42-45 mm in width. After 66 days of incubation, half of the eggs deteriorated; the remaining seven eggs began to shrink and apparently were loosing fluid. When opened, all contained dead fetuses.

Eastern Carpet Python

Morelia spilota mcdowelli (Wells & Wellington, 1984)

The Eastern Carpet Python ranges throughout eastern Queensland and parts of northeast New South Wales (Fig. 89). Considered the largest of the Carpet Pythons, specimens measuring 12.5 feet (3.81 m) and 14 feet (4.27 m) in length have been documented.[22] The average length, however, is about 8 feet (2.5 m). Specimens from southeast Queensland tend to be the largest.[13] Color and pattern consist of an intricate array of shades of brown, grey, and green. While some specimens are attractively marked, others may be drab (Fig. 17, 65,

67). Neonates from southern Queensland and northern New South Wales are red-orange, and sometimes exhibit red-brown banding. As they mature, however, these bright colors fade to muddy, brown-green hues. In captivity, this variation has been mistaken for both the Central Carpet Python and the Northern Carpet Python. This taxon can be differentiated from either subspecies based on color.

A variety of habitats are used, and specimens are frequently encountered along tree-lined rivers and creeks in rainforest, and among rocky hills adjacent to water (Figs. 30, 66, 73). These Carpet Pythons are primarily nocturnal, and often are seen crossing roads at night.

The Eastern Carpet Python preys on birds, flying foxes, rats, opossums, wallabies, and feral cats. They are notorious for entering yards, barns, and aviaries in search of prey. Specimens are periodically donated to the Queensland Reptile and Fauna Park. Individuals housed outdoors at the park have ingested items such as kangaroo skin wallets accidentally dropped into their enclosure.[14] Covacevich & Couper (1992) reported juveniles to feed on White's Tree Frogs (*Litoria* [*Pelodryas*] *caerulea*), and Williams (1992b) relates a second-hand account of two juveniles preying on invertebrates.

The average clutch size is 25 eggs, but clutches of 50 eggs are documented. Barker & Barker (1994b) reported 28 clutches of 12-48 eggs. Further, they found hatchlings to emerge at 49-64 days after oviposition. Neonates are about 16 inches (43 cm) in length.

Inland Rivers Carpet Python

Morelia spilota metcalfei (Wells & Wellington, 1985)

The Inland Rivers Carpet Python inhabits the Murray-Darling River system in Victoria, New South Wales, Queensland, South Australia, and adjacent areas (Fig. 89). Adults can reach 7-8 feet (2.15-2.46 m) in length, but average length is slightly smaller. These pythons have patches of red and grey-white on the dorsum. Some specimens have a blue-grey appearance, and those from New South Wales, South Australia, and Victoria, often show various shades of grey and black

(Barker & Barker, 1994b).[22] Ehmann (1992) reported specimens from the Eyre Peninsula and Flinders Range, South Australia as "pale reddish brown with broad transverse distinctive buff (black-edged) patches on the back and a very prominent wide, pale lateral zone for about a third of their body." In the northern part of the range, the red color is more prominent. Specimens from the northern extensions of the Murray Valley and Lake Eyre drainage basins, including those from Goyders Lagoon, South Australia, and Birdsville, Queensland, have pronounced, washed out patches of red and brown (Fig. 68). Although their pattern is similar to other Inland Rivers Carpet Pythons, they tend to be much paler in color, and are among the most beautiful of the Carpet Pythons.[10] A specimen bearing a strong resemblance to those populations was collected in Tennant Creek, Northern Territory, and was identified erroneously as a Central Carpet Python in Ross & Marzec (1990) (p. 142). Although this specimen may have been a transplant, there are anecdotal reports of the Inland Rivers Carpet Python from watercourses north of Birdsville and south of Mount Isa, in western Queensland.[10] The variation present in the northern distribution of the Inland Rivers Carpet Python remains undocumented.

This python inhabits the shallow meandering waterways and *Eucalyptus* lined river flats common throughout the inland wetlands and reedy marshes of this region (Fig. 69). Specimens also are found in areas of salt bush and occasional clump of mulga.[6] Shine (1993) states that rabbit warrens and rock outcrops are common refugia. Presumably, its diet is similar to that of other Carpet Pythons, with adults feeding primarily on mammals and birds, and juveniles on lizards for the first year of life.

The Inland Rivers Carpet Python is occasionally brought back to Melbourne by campers, but remains poorly studied both in captivity and in nature. It has been speculated that due to agricultural development, it is declining in parts of its habitat; but, in general, populations appear to remain stable (Wells & Wellington, 1985; Coventry & Robertson, 1991; Shine, 1993; Barker & Barker, 1994b).

Barker & Barker (1994b) reported a clutch of 21 eggs laid in November, with 19 neonates measuring about 19 inches (48.9 cm) in length. Incubation required 51-54 days. A gravid female from north-central Victoria measuring slightly over 5 feet (1.6 m) was observed in the early morning on 24 December 1994, crawling under a large boulder, presumably after basking in the sun. Careful palpation of an abdominal mass revealed 12 nearly-formed eggs. Further trips to the area failed to reveal the female or any neonates.[6]

Northern Carpet Python

Morelia spilota variegata (Gray, 1842)

The Northern Carpet Python is found throughout northern Australia, and has a similar distribution as the Northern Brown Python (Fig. 89). Adults are banded in beautiful shades of red, orange, black, and grey-white. They can attain a length of 8 feet (2.46 m), but the average length is about 5-6 feet (1.53-1.85 m) (Fig. 70). Specimens from the Kimberleys, Western Australia often display a more heavily banded pattern than those from the Northern Territory and often have a chained appearance. At all ages, the Northern Carpet Python resembles populations of New Guinea Carpet Pythons from the southern Irian Jaya-Papua New Guinea border (Barker & Barker, 1995c).

The habitat of this python is quite variable (Figs. 7, 52, 54, 71). In the Kimberleys, specimens inhabit areas of sandstone with closely bordering vine forest, and in the Northern Territory they are found mainly in densely forested areas alongside floodplains or roads.

Little is known on the natural history of this python. Gavin Bedford, of the Northern Territory University, recently has implanted radio transmitters in several specimens, and found their hunting behavior to closely resemble that of the Diamond Python. Specimens were found to ambush their prey in a favorable site for several days at a time. Their prey is similar to that of other Carpet Pythons, with juveniles feeding on lizards and adults preying on a variety of birds and mammals.

The reproductive biology of the Northern Carpet Python is poorly understood. Barker & Barker (1994b) reported male-male combat with the potential for biting in captive specimens. Kortlang (1991) documented two clutches, (11 and 12 eggs) laid in October and November, respectively. Neonates emerging from these clutches averaged 21 inches (55.8 cm) in length. Hatchlings were pink to brick-red and had a muted pattern, which slowly changed as they attained maturity (Fig. 20).

Interpreting Relationships in the *Morelia spilota* Group

Relationships among members of the *Morelia spilota* group have been studied for over two centuries. First described as *Coluber Arges* by Linnaeus (1758), the Carpet Pythons have since taken on an impressive number of both scientific and common names (*see* **A Checklist of Australian Pythons**). The Carpet Python, described as *Morelia spilota* by Underwood & Stimson (1990), is currently partitioned into seven subspecies or races. These include the Diamond Python, the Central Carpet Python, the Jungle Carpet Python, the Southwestern Carpet Python, the Eastern Carpet Python, the Inland Rivers Carpet Python, and the Northern Carpet Python.

Since the publication of Cogger (1992), three subspecies of the Carpet Python (*M. s. cheynei, M. s. mcdowelli,* and *M. s. metcalfei*) have been described. These names, first proposed as full species by Wells and Wellington (1984, 1985), were informally changed to subspecific status by Barker & Barker (1994b). Prior to this revision, *Morelia spilota variegata* (holotype locality: Port Essington, Northern Territory) was assigned to specimens from New South Wales, the Northern Territory (excluding *M. s. bredli*), Queensland, South Australia (excluding *M. s. imbricata*), Victoria, Western Australia (excluding *M. s. imbricata*) and extralimital localities, such as the Indonesian and Papuan populations. Thus, the Jungle Carpet Python, Eastern Carpet Python, Inland Rivers Carpet Python, Northern Carpet Python, and New Guinea Carpet Python were all considered regional variants of *M. s. variegata*. Not surprisingly, this arrangement is

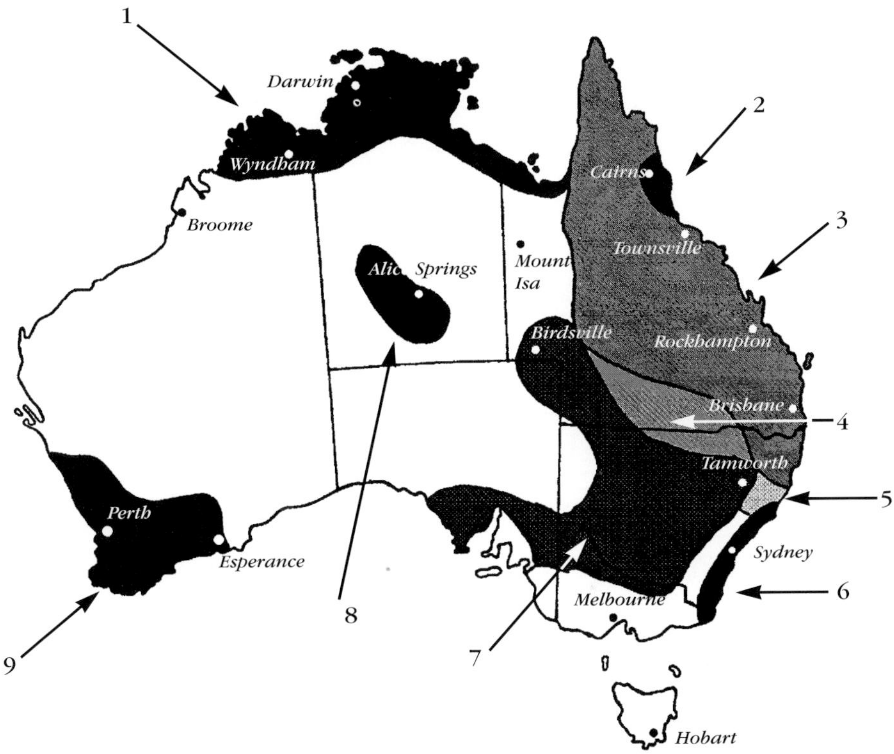

1. Northern Carpet Python
2. Jungle Carpet Python
3. Eastern Carpet Python
4. Overlap of the Eastern Carpet Python and the Inland Rivers Carpet Python
5. Intergredation area of the Diamond Python and Eastern Carpet Python
6. Diamond Python
7. Inland Rivers Carpet Python
8 Central Carpet Python
9. Southwestern Carpet Python

Figure 89. Distribution of the Carpet Python in Australia.

confusing and not biologically coherent.

Wells & Wellington published a revision of the Carpet Python complex in two papers (1984; 1985). These publications, however, are rarely cited by the academic community. In various ways Wells & Wellington (1984; 1985) went against accepted, standard scientific protocol, but the International Commission on Zoological Nomenclature refused to rule on an application to suppress their work. Kluge (1993) and Barker & Barker (1994b) have recognized their work where appropriate. Wells & Wellington (1984; 1985) were not challenged immediately, nor were new names proposed for the new taxa. Consequently, Barker & Barker (1994b) proposed three new combinations: *M. s. cheynei, M. s. mcdowelli,* and *M. s. metcalfei.* This arrangement has its difficulties, some mentioned earlier, but for the purpose of this book, these combinations aid in identification and permit discussion of the various Carpet Python populations. As indicated by Barker & Barker (1994b), the taxa *M. s. cheynei, M. s. mcdowelli,* and *M. s. metcalfe*i all correspond to regional forms frequently referred to in the literature by authors who did not accept the work of Wells & Wellington (Wilson & Knowles, 1988; Hoser, 1989; Ehmann, 1992).

The taxonomic view of Barker & Barker (1994b) simplifies the identification of the Carpet Pythons by relying on characters of color and pattern; morphological characters, such as scale counts, tend to strongly overlap in this group. Fyfe (*pers. comm.*), who has had extensive experience with *M. spilota* writes:

> I have had records of over 200 separate specimens in the *spilota* complex, including nearly 30 *spilota spilota*. Between subspecies, most of the scale counts overlap to some degree. There is no single character that will separate out or diagnose any of the proposed *variegata* populations (e.g., *cheynei, mcdowelli,* and *metcalfei),* making it impossible to diagnose any *M. spilota* by taxonomic key.

The results of unpublished data on Carpet Pythons may provide additional insights. Greg Mengden (*pers. comm.* in Montgomery, 1992) found that electrophoretic data on the "jungle phase" of the

Carpet Python is distinct from other members of the *Morelia spilota* group. The Jungle Carpet Python appears to be distinctive, as *M. s. bredli*, and possibly *M. s. spilota*. Steve Donnellan of the South Australian Museum, also has conducted protein electrophoresis in all proposed taxa of *M. spilota*. Preliminary results indicate that, on average, consistent genetic differences exist between the Southwestern Carpet Python and all other *M. spilota*.[8] No further differences, however, were detected between other populations of *M. spilota*. Populations of the Inland Rivers Carpet Python from the Eyre Peninsula, Flinders Range, and the lower Murray Valley drainage were found to be intermediate between the Southwestern Carpet Python and all other *M. spilota*. Also, protein analysis of the Southwestern Carpet Python from St. Francis Island, located off the coast of South Australia and along the Great Australian Bight, was found to be similar to mainland populations of the Southwestern Carpet Python. Investigation of *M. spilota* is in progress at the South Australian Museum with plans of DNA studies.

The Central Carpet Python appears to be geographically separated from both the Inland Rivers Carpet Python and the Northern Carpet Python (Fig. 89), but Ingram & Raven (1991) record a specimen from Birdsville, southwestern Queensland. It is possible, however, that this was a released animal, but if additional specimens are documented, this would bridge the gap between the Central Carpet Python and the Inland Rivers Carpet Python. Likewise, the Granites, in the Tanami Desert about 500 km North of Alice Springs, are considered northernmost locality for the Central Carpet Python (Gow, 1981; Fyfe, 1994). No museum specimens are available, but there are anecdotal reports of Carpet Pythons from the Gardner Range, on the Northern Territory-Western Australia border.[10] Additional specimens from this region may link the distribution of the Central Carpet Python and the Northern Carpet Python.

Moving southward to the Great Australian Bight, Schwaner et. al. (1988) reported Southwestern Carpet Pythons from islands off the South Australian coast (Fig.3). The taxonomic identification of this population has varied between authors (Schwaner et. al., 1988; Ehmann, 1992; Barker & Barker, 1994b), but I agree with Schwaner et.

al., (1988) and Barker & Barker (1994b) that, in spite of the geographic location of these populations, they should be recognized as Southwestern Carpet Pythons. It has been hypothesized that in the past, St. Francis specimens may have formed the link between the Southwestern Carpet Python and the Inland Rivers Carpet Python. Carpet Pythons from Sceale Bay, near Ceduna, western Eyre Peninsula, South Australia strongly resemble St. Francis Island specimens.[10]

Along the coast of the Great Australian Bight and throughout the Nullarbor Plains, the taxonomic status of the Carpet Python is uncertain and poorly documented. Hoser (1989) and Cogger (1992) have included this region in the range of *M. spilota*, but I have been unable to receive information on these populations. Perhaps, in the past, Carpet Pythons were found throughout this area.[10] At present Carpet Pythons do not appear to have a continuous distribution along the Nullarbor, which is a flat and treeless region. It is possible, however, that isolated populations exist in the scattered limestone cave systems.[10] Until further documentation becomes available, I have tentatively included these populations under the Southwestern Carpet Python.

In coastal northern Queensland, the Jungle Carpet Python is sympatric with the Eastern Carpet Python. The latter is apparently far less common.[22] Throughout the open woodlands of the Atherton Tablelands, the Eastern Carpet Python usually is dark brown, black, grey, or olive (Montgomery, 1992) (Fig. 67). Intergrade specimens have not been reported. Clifford Frith and Dawn Frith (*pers. comm.*), who live in Malanda, Atherton Tablelands, have not observed these forms to interbreed or to have found specimens with intermediate characteristics. Barker & Barker (1994b) indicated five characters that differentiate the Jungle Carpet Python from the Eastern Carpet Python. Importantly, they state that the Jungle Carpet Python typically has a distinct black line through the eye and a sharply defined head pattern, characters that the Eastern Carpet Python does not share.

There is some uncertainty as to whether the name *M. s. mcdowelli* is correct for the Eastern Carpet Python. The name *Python spilotes macropsila* was proposed for this population by Werner (1909), type locality of "Australia," but unfortunately the holotype was lost. This subspecies was recognized by Hoser (1989). If the holotype eventually is found, *M. s. macropsila* may replace *M. s. mcdowelli* as the subspecific name for the Eastern Carpet Python.

Eastern Carpet Pythons from northern Queensland, particularly from the Cape York Peninsula, are poorly known and there are few specimens in museum collections (Barker & Barker, 1994b) (Fig. 65). Carpet Pythons from Cape York may not have a continuous distribution.[17] A specimen from Bamaga, on the tip of the Cape York Peninsula, closely resembles the Northern Carpet Python in its uniform banding and head coloration.[10] Barker & Barker (1994b) stated that *M. spilota* from the Cape York Peninsula may be a distinct form.

There are two areas where populations of *M. spilota* overlap (Fig. 89). Weigel (1990) and Barker & Barker (1994b) report that in northern New South Wales, between Taree and Coff's Harbour, intergradation between the Diamond Python and the Eastern Carpet Python occurs. Weigel (1990) noted that adult specimens from this region appear more brown and have less spotting, while neonates are red-orange, similar to the Eastern Carpet Python from northern New South Wales and Southern Queensland. Specimens found in this region tend to look like intermediates.[10] A second area of overlap occurs in central northern New South Wales and central southern Queensland, between the Eastern Carpet Python and the Inland Rivers Carpet Python. I could not find information on this population. Thus, based on lack of data, I have mapped the intergrade distribution generously, but consider it to be tentative.

Green Python

Morelia viridis (Schlegel, 1872)

The Green Python is Australia's most arboreal python. It is restricted to the eastern rainforests of the Cape York Peninsula, Queensland (Fig. 90), and only known from several localities: the Iron Range National Park, the Mcllwraith Range, and the Sir Walter Thompson Range. There also are verbal reports of this python from the rainforests of the Lockerbie

Scrub region, on the northern tip of Cape York, but these are yet to be verified.[17] Australian specimens generally have a broken or complete series of white to yellow markings present along their vertebral line, a trait also shown by individuals from several Indonesian and Papuan localities (Fig. 72). Specimens from both Australia and abroad may have blue and or yellow markings centered on the vertebral line. As adults, Green Pythons are 4.5-5.5 feet (1.38-1.69 m) in length.

This species inhabits low trees and hollows, shrubs, and garden vegetation (Fig. 73).[5] It is primarily nocturnal, and during the day rests in a coiled position, usually on a branch. At night, these pythons leave their arboreal roost and search for food.[9] The diet of adults consists primarily of birds and mammals, while hatchlings feed on treefrogs and small lizards.[22] An ornithologist involved in a radiotelemetry project reported an adult Green Python to devour a Rifle Bird (*Ptilorus paradiseus*) high in the tree canopy.[14] According to Steve Irwin of the Queensland Reptile and Fauna Park, the principal diet of adult Green Pythons consists of the Cape York Rat (*Rattus leucopus*). These rats nest in the trees, and descend at night to gather food in the grasses and shrubs at the base of tree thickets. Green Pythons often inhabit these clumps of grass and shrubbery.

In Australia, Green Pythons are notoriously difficult to locate in the field, and little is known of their natural history, including reproduction. It is unknown in nature whether or not specimens reproduce seasonally or aseasonally. In captivity, female specimens from Irian Jaya, Papua New Guinea, and surrounding areas have been observed to change color from green to turquoise-blue when gravid, an observation yet to be reported in specimens from Australia. The Green Python lays clutches of 12-22 eggs, and hatching occurs in 50-55 days. Neonates are 11-12 inches (28.2-30.76 cm) in length.[22]

Neonates are bright yellow, red, or maroon. The yellow neonates may change color quickly, sometimes in as little as two weeks, although the time can be quite variable between specimens. Some specimens may continue changing their color for several years. One hatchling Australian specimen presumably changed from brilliant yellow to a drab olive green

Figure 90. Distribution of the Green Python in Australia.

overnight. Two weeks later, and without shedding, this specimen showed its adult coloration.[9] For red and maroon babies, color ontogeny appears more complicated, generally taking up to six months to be completed.

Color ontogeny in reptiles is poorly understood, and the purpose of color ontogeny in Green Pythons remains unexplained. It is interesting to note, however, that inhabitants of Papua New Guinea have encountered yellow neonates in thickets of "Finisterre" (*Schizostachyum* spp.), a species of bamboo prominent throughout the river valleys of Irian Jaya and Papua New Guinea. When mature, this plant is noted for its hollow culms, which are generally 8-10 cm in diameter, and its bright yellow color. This may be evidence of protective coloration in the yellow neonates. This does not take into account, however, the presence of red or maroon neonates.

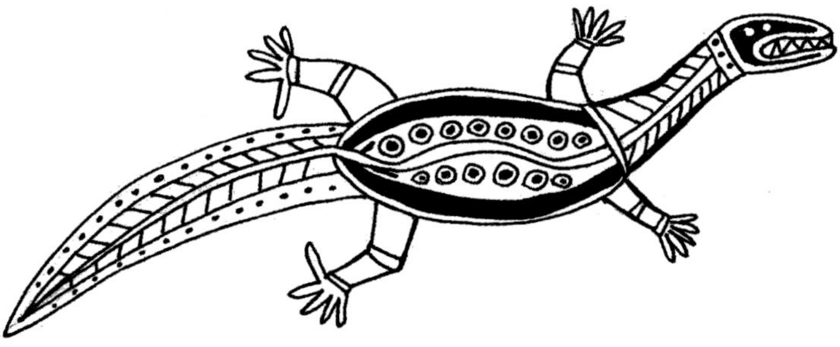

Epilogue

"Good natural history is permanently archived, so that it is available for future users...[and] serves as a factual basis for the life sciences, for nature conservation, and for environmental appreciation."

-Harry W. Greene, PhD

Brian Kend has assembled and presented a large amount of information on Australian pythons, and throughout the pages of this book, has stressed three main areas: systematics and taxonomy, natural history, and conservation.

In my opinion, research on the systematics and taxonomy of pythons is ongoing and in a healthy state of inspection. Recent works by Dr. Arnold Kluge are important in that the method he used to reconstruct phylogenies (phylogenetic systematics) is scientific, relevant to critical testing, and forms the basis for developing meaningful taxonomies.

The late Dr. Laurence M. Klauber, an expert on rattlesnakes, said "Who, in wandering about the hills in search of rattlesnakes, has not wished for a special Geiger counter responsive to some unknown *Crotalus* emanation, whereby he might locate every rattler in the area..." Today, that wishful thinking has become, in part, a reality. Research on snakes in nature has undergone a revolution in the past 20 years, due almost entirely to technological innovations in the fields of radiotelemetry and electronic miniaturization. The use of radiotelemetry in the study of Australian pythons and other snakes is rapidly growing, as seen in the published works of Dr. Richard Shine and colleagues cited herein.

Kend has assembled a wide range of anecdotes from both amateur and professional biologists, reminiscent of the style that Klauber used in his significant volumes on rattlesnakes. Criticized for being unscientific, anecdotes hold great utility and often form the basis for further research, and they serve to archive information on populations that otherwise might be lost. When explaining the importance of anecdotal data, I often tell my students the famous quote by Daniel J. Boorstin, Library of Congress Emeritus, "*You must collect things for reasons you don't yet understand.*" Indeed, Kend's book does a tremendous service in presenting these anecdotes of the natural history of Australian pythons.

The conservation of Australian pythons and all snakes is critical for the health of our planet. For those who do not appeciate snakes, there must be an attempt to abandon deep-rooted fears and prejudices, and to acknowledge their importance in the dynamics of ecosystems. In many ways, the manner in which we educate the public is the key to their survival. If this book has brought the lives of Australian pythons closer to your own, Brian has been successful.

Gordon W. Schuett, PhD
Department of Life Sciences
Arizona State University West
August, 1997

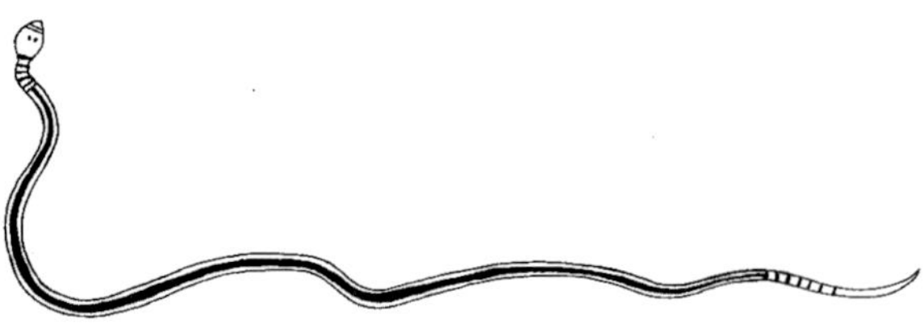

Personal Communication

I gratefully acknowledge the following people for their personal communication (*pers. comm.*) of information used in this book. All superscripts (parenthetical numbers) identify the specific individuals or source.

Corey Blanc (**1**)
Rob Bredl (**2**)
Brian Bush (**3**)
Neil Charles (**4**)
Harold Cogger (**5**)
Peter Comber (**6**)
Dale DeNardo (**7**)
Steve Donnellan (**8**)
Ray Field (**9**)
Greg Fyfe (**10**)
Graeme Gow (**11**)
Greg Harold (**12**)
Raymond Hoser (**13**)
Steve Irwin (**14**)
Casey Lazik (**15**)
Matthew Moyle (**16**)
Lyall Naylor (**17**)
Terry Schwaner (**18**)
Richard Shine (**19**)
Laurie Smith (**20**)
John Weigel (**21**)
Bob Withey (**22**)

Glossary

Ambient: External or outside.
Arboreal: Dwelling in trees.
Aseasonal: No specific season or period.
Biennial: Every two years.
Boid: A taxonomic category referring to species of snakes called pythons and boas.
Caudally: Referring to or the area near the tail.
Cloaca: Chamber into which the reproductive, intestinal, and urinary ducts open. The external opening of the cloaca often is referred to as the vent.
Cloacal Gaping: An indication of female receptivity to male courtship. The female elevates her tail, lowers her vent, and exposes her cloaca.

Colubrid: A lineage of mostly non-venomous snakes.
Crepuscular: Active occurs during the evening and early morning.
Cryptic: Hidden or difficult to see; background matching.
Dingo: Australian Wild Dog (*Canis familiaris dingo*).
Diurnal: Active during the day.
Dorsal: Back region.
Ecdysis: Shedding the skin.
Ecology: A branch of biology dealing with the relation of living organisms to their environment.
Ectotherms: Organisms that rely on external heat to carry out most metabolic processes.
Endotherms: Organisms that uses metabolically-derived heat.
Extralimital: Beyond the limits of a country or district.
Fossorial: Adapted for digging or burrowing; living in subterranean refuges.
Gekkonid: The taxonomic category referring to species of lizards called geckos.
Gestation: The period of time in which embryos develop within the body of females prior to egg laying or live birth.
Goanna: The Australian term for monitor lizards, genus *Varanus*.
Gravid: Pregnant; usually restricted to oviparous taxa.
Hemipenis: Bi-lobed copulatory organ in male squamates (snakes and lizards).
Herpetoculture: Amphibian and reptile maintenance and propagation.
Herpetofauna: Amphibians and reptiles.
Herpetologist: One (usually an expert) interested in herpetology, the study of amphibians and reptiles.
Hybridization: Cross mating and reproduction between different species.
Insular: Island dwelling.
Intergradation: Reproduction between subspecies or geographic variants (*see* Hybridization).
Labials: Scales bordering the mouth.
Laterite: Red soil.
Male Aggregation: Two or more males in close proximity to a single female.

Microhabitat: Specific type of place in an animal's habitat, (e.g., burrow).
Mulga: Various small Australian acacias, yielding a hard, durable wood.
Nocturnal: Active at night.
Ontogenetic: The development (life-cycle) of an individual organism, from conception to death.
Ophiophagous: To eat snakes.
Oviparous: Egg laying.
Oviposition: The process of egg-laying.
Ovulation: Release of mature ova into the oviducts.
Phylogeny: The genealogical history of species.
Physiology: The branch of science concerned with the integrative functions of living organisms and their parts.
Pilbara: A desert region north of Perth and south of the Kimberleys in Western Australia.
Radiotelemetry: The use of electronic radiotransmitters often used in monitoring movements of free-ranging animals.
Scincid: The taxonomic category referring to lineages of lizards called skinks.
Seasonal: Specific season or period.
Slough: Shed-skin (*see* Ecdysis).
Speciose: A taxonomic group having many species.
Synonymy: List of different scientific names referring to a single species or group.
Subordinate: Lower in rank or status.
Taxon (pl: **Taxa**): Category in a system of classification.
Terrestrial: Living on land.
Thermoregulation: The capacity to modify body temperature.
Top End: Northern part of the Northern Territory.
Triennial: Every three years.
Varanid: A taxonomic category referring to a lineage of lizards called monitors (*see* Goanna).
Ventral: Bottom plane (surface) of an organism.
Vestige: Morphological or behavioral characters no longer fully developed and may not have a use; these characters likely had a different function at an earlier evolutionary stage.

Literature Cited

Barbour, T. 1914. On some Australasia reptiles. *Proceedings of the Biological Society of Washington*, **27**, 201-206.

Barker, D.G., J.B. Murphy & K.W. Smith. 1979. Social behavior in a captive group of Indian pythons, *Python molurus* (Serpentes, Boidae) with formation of a linear social hierarchy. *Copeia*, **1979** 466-471.

Barker, D. 1984. Maintaince and reproduction of black-headed pythons at the Dallas Zoo. In: *8th Annual Reptile Symposium on Captive Propagation and Husbandry* (Ed. by R. Hahn). Thurmont, Maryland: Zoological Consortium, Inc.

Barker, D.G. & T.M. Barker. 1994a. The African giants. *The Vivarium*, **6**(1), 18-21.

Barker, D.G. & T.M. Barker. 1994b. *Pythons of the World. Vol. 1, Australia.* Lakeside, California: Advanced Vivarium Systems.

Barker, D.G. & T.M. Barker. 1995a. The southern stars of Australia: Australia's small pythons. *The Vivarium*, **6**(4), 30-35.

Barker, D.G. & T.M. Barker. 1995b. The mechanics of python reproduction. *The Vivarium*, **6**(5), 30-33.

Barker, D.G. & T.M. Barker. 1995c. A new python in captivity from New Guinea, the New Guinea carpet python and the Sawu python, a correct common name. *The Vivarium*, **6**(6), 30-33.

Barnett, B. 1994. The Amethystine python (*Morelia amethistina*). captive keeping, reproduction, and growth. *Litteratura Serpentium*, **13**, 184-195.

Bedford, G.S. & T. Sullivan. 1993. Observations of intra- and interspecific aggression and possible mating hierarchy in the Northern children's python (*Liasis childreni*). *Herpetofauna*, **23**, 10-12.

Boos, H.E. 1979. Some breeding records of Australian pythons. In: *International Zoo Yearbook*, Vol. 19 (Ed. by P.S. Olney), pp. 87-89. London: Zoological Society of London.

Boulenger, G. 1893. *Catalogue of the Snakes in the British Museum (Natural History)*. London: British Museum.

Brongersma, L.D. 1953. Notes on New Guinean reptiles and amphibians. *Proceedings Koninklijke Nederlandsche Akademie van Wetenschappen*, **59**, 290-300.

Brongersma, L.D. 1956. On two species of boid snakes from the Lesser Sunda Islands. *Proceedings Koninklijke Nederlandsche Akademie van Wetenschappen*, **59**, 290-300.

Bush, B. 1981. *Reptiles of the Kaloorlie-Esperance Region*. Perth, Australia: Vanguard Press.

Bush, B. 1988. An unsuccessful breeding record for the Western Australian carpet python, *Morelia spilota imbricata*. *Herpetofauna*, **18**, 30-31.

Charles, N. 1988. Captive reproduction of seven species of Australian snakes in the families Elapidae and Boidae. In: *10th and 11th International Herpetological Symposia on Captive Propagation and Husbandry* (Ed. by R. Hahn). Thurmont, Maryland: Zoological Consortium, Inc.

Charles, N., R. Field & R. Shine. 1985. Notes on the reproductive biology of Australian pythons, genera *Aspidites*, *Liasis*, and *Morelia*. *Herpetological Review*, **16**, 25-48.

Charles, N. & P. Wilson. 1985. A caesarian operation on an Oenpelli Python. *Thylacinus*, **10**, 108-113.

Cogger, H.G. 1992. *Reptiles and Amphibians of Australia*. Sydney, Australia:AH and AW Reed Pty. Ltd., Publ.

Cogger, H.G. & A. Holmes. 1960. Thermoregulatory behavior in a specimen of *Morelia spilotes variegata* Gray (Serpentes, Boidae). *Proceedings of the Linnean Society* (New South Wales), **85**, 328-333.

Cogger, H.G. & D.A. Lindner. 1974. Frogs & reptiles. In: *Fauna Survey of the Port Essington District, Cobourg Peninsula, Northern Territory of Australia* (Eds. H.J. Frith & J.H. Calaby), pp. 63-108. Sydney, Australia: Commonwealth Scientific and Industrial Research Organization.

Cogger, H.G., E.E. Cameron, & H.M. Cogger. 1983. *Zoological Catalogue of Australia. Vol. 1. Amphibia and Reptiles*. Canberra, Australia: Australian Govt. Publ. Serv.

Cogger, H.G., E. Cameron, R. Sadlier, & P. Eggler. 1993. Species recovery outline: Western Australian carpet python. *The Action Plan for Australian Reptiles*. Australian Nature Conservancy Agency. Australian Museum, Sydney. **December 1993**, 134-137.

Covacevich, J. & P. Couper. 1992. The carpet python, *Morelia spilota* (Lacepede), another unsuccessful predator of the cane toad *Bufo marinus* (Linnaeus), in Australia. In: *Contributions in Herpetology* (Ed. by P.D. Strimple & J.L. Strimple), pp. 57-59. Greater Cincinnati Herpetological Society: Cincinnati, Ohio

Coventry, A.J. & P. Robertson.1991. *The Snakes of Victoria*. Melbourne, Australia: Department of Conservation & Environment.

Dumeril, A.M.C. & G. Bibron. 1844. *Erpetologie generale ou histoire naturelle complete des reptiles. Vol. 6*. Libraire Encyclopedique de Roret, Paris.

Ehmann, H. (Ed.). 1992. *The Encyclopedia of Australian Animals: Reptiles*. New South Wales, Australia: Angus & Robertson.

Ehmann, H. 1993. Family Boidae. In: *Fauna of Australia: Volume 2A Amphibia and Reptilia*, pp. 284-289. Canberra, Australian Capital Territory: Australian Government Publishing Service.

Ellis, T.M. & M.A. Chappell. 1987. Metabolism, temperature relations, maternal behavior, and reproductive energetics in the ball python (*Python regius*). *Journal of Comparative Physiology B* **157**:393-402.

Field, R. 1990. Developing techniques to breed pythons in captivity. *Herpetofauna*, **20**, 18-22.

Foster, R. 1993. Incubatory thermoregulation through clutch shifting by a Children's Python (*Liasis maculosus*). *South Australian Herpetological Society Newsletter*, **3**, 1.

Fyfe, G. & C. Harvey. 1981. Some observations on the woma (*Aspidites ramsayi*) in captivity. *Herpetofauna*, **13**, 23-25.

Fyfe, G. 1990. Notes on the central carpet python (*Morelia spilota bredli*). *Herpetofauna*, **20**, 11-14.

Fyfe, G. 1994. The central carpet python (*Morelia spilota bredli*) in the field and in captivity. *Monitor*, **6**, 7-12.

Gow, G.F. 1977. A new species of python from Arnhem Land. *The Australian Zoologist*, **19**, 133-139.

Gow, G.F. 1981. A new species of python from Central Australia. *Australian Journal of Herpetology*, **1**, 29-34.

Gow, G.F. 1989. *Graeme Gow's Complete Guide to Australian Snakes*. New South Wales, Australia: Angus and Robertson.

Gray, J. 1840. *Synopsis of the Contents of the British Museum*. 42nd Edn. London: British Museum.

Gray, J. 1842. Synopsis of the species of prehensile-tailed snakes, or family Boidae. In: *Zoological Miscellany* (Ed. by J. Gray) pp. 41-46. London: Treuttel, Wurtz, & Co.

Greer, G.C. 1994. Nest sites of royal pythons in West Africa. *Reptile and Amphibian Magazine*, **29**, 44-55.

Günther, A. 1879. Notice of a collection of reptiles from islands of Torres Straits. *Annals and Magazine of Natural History*, Ser. 5, **3**, 84-87.

Hardy, D. 1994. A re-evaluation of suffocation as the cause of death duing constriction by snakes. *Herpetological Review*, **25**, 45-47.

Hawkes, T., T. Jurgenson, & L. Telford. 1993. Tully Gorge field trip. *Chondro: Journal of the Cape York Herpetological Society*, **1**, 11-12.

Henderson, R.W. 1993. On the diets of some arboreal boids. *Herpetological Natural History*, **1**, 91-96.

Hingley, K.J., & N. Hemmings. 1993. Keeping and breeding the carpet python, *Morelia spilota ssp. Herptile*, **18**, 118-129.

Hoser, R.T. 1981a. Australian pythons, Part 1: Genera *Chondropython* and *Aspidites*. *Herptile*, **6**, 10-16.

Hoser, R.T. 1981b. Australian pythons, Part 2: The smaller *Liasis*. *Herptile*, **6**, 13-19.

Hoser, R.T. 1981c. Australian pythons, Part 3: The larger *Liasis*. *Herptile*, **6**, 3-12.

Hoser, R.T. 1982. Australian pythons, Part 4: Genus *Morelia* and *Python carinatus*, followed by discussions on the taxonomy and evolution of Australasian pythons. *Herptile*, **7**, 2-17.

Hoser, R.T. 1989. *Australian Reptiles and Frogs*. Sydney, Australia: Pierson & Company.

Hoser, R.T. 1992. Search for the ant-hill python, *Bothrochilus perthensis*. *Litteratura Serpentium*, **12**, 13-19.

Hoser, R.T. 1993. Children's pythons and lookalikes: The *childreni* Complex. *Reptilian Magazine*, **1**, 10-15.

Hoser, R.T. 1995. The ant-hill python. *Reptiles*, **18**, 10-16.

Hubrecht, A. 1879. On a new genus and species of Pythonidae from Salawatti. *Notes from the Leyden Museum*, **1**, 14-15.

Ingram, G.J. & R.J. Raven. 1991. *An Atlas of Queensland's Frogs, Reptiles, Birds, and Mammals*. Brisbane, Australia: Trustees, Queensland Museum.

Kend, B. 1992. The small pythons of Australia. In: *Unedited Proceedings of the 16th International Herpetological Symposium on Captive Propagation and Husbandry*, pp. 154-159. International Herpetological Symposium, Inc.

Kinghorn, J. 1964. *Snakes of Australia*. New South Wales, Australia: Angus and Robertson.

Kluge, A.G. 1991. Boine phylogeny and research cycles. *Miscellaneous Publications Museum of Zoology University of Michigan*, **178**, 1-58.

Kluge, A. 1993. Aspidites and the phylogeny of pythonine snakes. *Records of the Australian Museum,* **19**, 1-77.

Kortlang, S. 1991. Husbandry and reproduction of the Northern Territory/Kimberley form carpet python, *Morelia spilota variegata* (Gray, 1842). *Monitor (Bulletin of the Victorian Herpetological Society),* **3**, 51-60.

Krauss, P. 1992. Husbandry and captive breeding of the Oenpelli Python. *Thylacinus*, **17**, 2-6.

Krefft, G. 1864. Description of *Aspidiotes melanocephalus*, a new snake from Port Denison, N.E. Australia. In: *Proceedings of the Zoological Society of London,* pp. 20-22. London, Great Britain.

Lacepede, B.G. 1804. Memoire sur plusieurs animaux de la Nouvelle-Holland dont la description n'a pas encore ete publiee. *Annals du Museum National d'Histoire Naturelle*, Paris, **4**, 184-211.

Lethbridge, P. 1993. An unusual feeding account of an Amethystine Python *Morelia amethistina*. *Chondro: Journal of the Cape York Herpetological Society*, **1**, 22.

Linnaeus, C. 1758. *Systema Naturae per regna tria naturae, secundum classes, ordines, genera, species, cum characteribus, differentiis, synonymis, locis*. Vol. 1. Edn 10. Holmiae: Laurentii.

Longman, H.A. 1913. Herpetological notes I. Systematic. Including the description of one new species. Part II. Ethological. *Memoirs of the Queensland Museum*, **2**, 39-45.

Loveridge, A. 1948. Australian Reptiles in the Museum of Comparative Zoology, Cambridge, Massachusetts. *Bulletin of the Museum of Comparative Zoology*, **77**, 243-383.

Macleay, W. 1882. Description of two new species of snakes. *Proceeds of the Linnean Society of New South Wales*, **6**, 811-813.

Macleay, W. 1885. On some Reptilia lately received from the Herbert River District, Queensland. *Proceeds of the Linnean Society of New South Wales*, **10**, 64-68.

Madsen, T. & R. Shine. 1996a. Determinants of reproductive output in female water pythons (*Liasis fuscus*: Pythonidae). *Herpetologica*, **52**, 146-159.

Madsen, T. & R. Shine. 1996b. Seasonal migration of predators and prey--a study of pythons and rats in tropical Australia. *Ecology*, **77**, 149-156.

McDowell, S. 1975. A cataloque of the snakes of New Guinea and the Solomons, with special reference to those in the Bernice P. Bishop Museum. Part II. Anilioidea and Pythoninae. *Journal of Herpetology*, **9**, 1-79.

McDowell, S. 1979. A catalogue of the snakes of New Guinea and the Solomons, with special reference to those in the Bernice P. Bishop Museum. Part 3. Boinae and Acrochordoidea (Reptilia, Serpentes). *Journal of Herpetology*, **13**, 1-92.

Merrem, B. 1820. *Versuch eines Systems der Amphibien*. Marburg: J.C. Krieger.

Meyer, A.B. 1874. Eine Mitteilung von Hrn. Dr. Adolf Bernhard Meyer uber die von ihm auf Neu-Guinea den Inseln Jobi, Mysore und Mafoor im Jahr 1873 gesammelten Amphibien. *Monatsberichte der Koniglich Preussischen Akademie der Wissenschaften zu Berlin*, **1874**, 128-140.

Montgomery, W.B. 1992. The “jungle phase” carpet python, *Morelia spilota*, husbandry and breeding. In: *Proceedings of the 15th International Herpetological Symposium on Captive Propagation and Husbandry* (Ed. by M. Uricheck), pp. 21-27. International Herpetological Symposium, Inc.

Müller, S. 1844. Bijdragen tot de kennis van Timor en eenige andere naburige Eilanden. Verhandelingen over de Natuurlijke Geschiedenis der Nederlandsche overzeesche bezittigen. *Land en Volkenkunde*, **7**, 193-224.

Müller, S. 1857. *Reizen en Onderzoekingen in den Indischen Archipel, gedaan op last der Nederlandsche Indische Regeering, tusschen de jaren 1828 en 1836*. Vol. 2. Amsterdam: Fred Muller.

Murphy, J.B., C.C. Carpenter, & J.C. Gillingham. 1978. Caudal luring in the green tree python, *Chondropython viridis* (Reptilla, Serpentes, Boidae). *Journal of Herpetology*, **12**, 117-119.

O'Brien, M. & L. Naylor 1987. Observational note on a black-headed Python. *Thylacinus*, **12**, 9.

Ogilby, J. 1891. Descriptions of three new Papuan snakes. *Records of the Australian Museum*, **1**, 192-194.

Pearson, D. 1993. Distribution, status, and conservation of pythons in Western Australia. In: *Herpetology in Australia: a Diverse Discipline* (Ed. by D. Lunney & D. Ayers), pp. 383-395. Sydney, Australia: Surrey Battey & Sons.

Peters. W. 1873. Uber ein neue Schildkrotenart, *Cinosternon effeldti* und einigre andere neue oder weniger bekannte Amphibien. *Monatsberichte der Konniglich Preussischen Akademie der Wissenschaften zu Berlin*, **1873**, 603-618.

Peters, W. 1876. *Aspidiotus/Aspidites. Monatsberichte der Koniglich Preussischen Akademie der Wissenschaften zu Berlin*, **1**, 914.

Peters, W. & G. Doria. 1878. Catalogo dei rettili e dei batraci raccolti da O. Beccari, L.M. D'Albertis e A.A. Bruijn nella sotto-regione Austro-malese. *Annali del Museo Civico di Storia Naturale di Genova*, **13**, 323-450.

Ross, R. & G. Marzec. 1990. *The Reproductive Husbandry of Pythons and Boas*. Stanford, California: The Institute for Herpetological Research.

Sauvage, H. 1878. Eassai sur la faune herpetologique de la Nouvelle-Guinee, suivi de la description de quelques especes nouvelles ou peu connues. *Bulletin des Sciences, par la Societe Philomathique de Paris*, **2**, 25-44.

Schlegel, H. 1872. *De Dierentuin Van het Koninklijk Zoologische "Natura Artis Magistra" te Amsterdam zoologisch door Prof. H. Schlegel, met historische herinneringen van P.H. Witkampf*. Gebr, Amsterdam.

Schneider, J.G. 1801. *Historiae amphibiorum naturalis et liturariae. Fasciculus secundus, continens crocodilos, scincos, chamaesauras, boas, pseudoboas, elapes, angues, amphisbaenas, et caecilias*. Jena (Fried. Frommann).

Schuett, G.W. 1997. Body size and agonistic experience affect fighting and mating success in male copperheads. *Animal Behaviour* (summer 1997).

Schuett, G.W., D.L. Clark, & F. Kraus. 1984. Feeding mimicry in the rattlesnake *Sistrurus catenatus*, with comments on the evolution of the rattle. *Animal Behaviour.*, **32**, 625-626.

Schwaner, T., M. Francis, & C. Harvey. 1988. Identification and conservation of carpet pythons (*Morelia spilota imbricata*) on St. Francis Island, South Australia. *Herpetofauna*, **18**, 13-18.

Shine, R. 1988. Parental care in reptiles. In: *Biology of the Reptilia. Volume 16, Ecology B.* (Ed. by C. Gans and R. Huey), pp. 605-694. New York: Wiley-Interscience.

Shine, R. & D.J. Slip. 1990. Biological aspects of the adaptive radiation of Australasian pythons (Serpentes: Boidae). *Herpetologica*, **46**, 283-290.

Shine, R. 1991. *Australian Snakes: A Natural History*. Sydney, Australia: Reed Books Pty. Ltd.

Shine, R. 1992. Ecological studies on Australian pythons. In: *Proceedings of the 15th International Herpetological Symposium on Captive Propagation and Husbandry* (Ed. by M. Uricheck), pp. 29-39. International Herpetological Symposium, Inc.

Shine, R. 1993. The biology and management of diamond pythons (*Morelia spilota spilota*) and carpet pythons (*Morelia spilota variegata*) in New South Wales. *Draft Species Management Report*. NSW National Parks and Wildlife Service.

Shine, R. 1994. Sexual size dimorphism in snakes revisited. *Copeia*, **1994**, 326-346.

Slip, D.J. & R. Shine. 1988a. Habitat use, movements, and activity patterns of free-ranging diamond pythons, *Morelia s. spilota* (Serpentes: Boidae): a radio-telemetric study. *Australian Wildlife Research*, **15**, 515-531.

Slip, D.J. & R. Shine. 1988b. Reptilian endothermy: a field study of thermoregulation by brooding diamond pythons. *Journal of Zoology*, **216**, 367-378.

Slip, D.J. & R. Shine. 1988c. Thermoregulation of free-ranging diamond pythons, *Morelia spilota* (Serpentes: Boidae). *Copeia*, **1988**, 984-995.

Slip, D.J. & R. Shine. 1988d. Feeding habits of the diamond python, *Morelia s. spilota*: ambush predation by a boid snake. *Journal of Herpetology*, **22**, 323-330.

Slip, D.J. & R. Shine. 1988e. Thermophilic response to feeding of the diamond python, *Morelia s. spilota* (Serpentes: Boidae). *Comparative Biochemistry and Physiology*, **89A**, 645-650.

Slip, D. & R. Shine. 1988f. The reproductive biology and mating system of the diamond python, *Morelia spilota* (Serpentes: Boidae). *Herpetologica*, **44**, 396-404.

Smith, L. 1981a. A revision of the python genera *Aspidites* and *Python* (Serpentes: Boidae) in western Australia. *Records of the Western Australian Museum*, **9**, 211-226.

Smith, L. 1981b. A revision of the *Liasis olivaceus* species-group (Serpentes: Boidae) in western Australia. *Records of the Western Australian Museum*, **9**, 227-233.

Smith, L. 1985. A revision of the *Liasis childreni* species group (Serpentes: Boidae). *Records of the Western Australian Museum*, **9**, 211-216.

Smith, M.J. & M. Plane. 1985. Pythonine snakes (Boidae) from the Miocene of Australia. *Bureau of Mineral Resources Journal of Australian Geology and Geophysics*, **9**, 191-195.

Stafford, P.J. 1986. *Pythons and Boas*. Neptune City: Tropical Fish Hobbyist, Inc.

Stimson, A.F. 1969. Liste der rezenten Amphibien und Reptilien. Boidae. *Das Tierreich*, **89**, 1-49.

Stimson, A.F. & S.B. McDowell. 1986. *Liasis Gray*, 1842 (Reptilia, Serpentes): proposed designation of *Liasis mackloti* Dumeril and Bibron, 1844 as Type species. *Bulletin of Zoological Nomenclature*, **43**, 330-334.

Storr, G.M, L.A. Smith & R.E. Johnstone. 1986. *Snakes of Western Australia*. The Western Australia Museum, Perth, Australia.

Strimple, P.D. 1992. Caudal luring: A discussion on the definition and application of the term. In: *Contributions in Herpetology* (Ed. by P.D. Strimple & J.L. Strimple), pp. 49-54. Greater Cincinnati Herpetological Society: Cincinnati, Ohio.

Stull, O.G. 1932. Five new species of the family Boidae. *Occasional Papers of the Boston Society of Natural History*, **8**, 25-30.

Stull, O.G. 1933. Two new subspecies of the family Boidae. *Occasional Papers of the Museum of Zoology. University of Michigan:* 1-4.

Stull, O.G. 1938. Three new subspecies of the family Boidae. *Occasional Papers of the Boston Society of Natural History*, **8**, 297-300.

Szokolay, S. 1983. *Climatic Data and Its Use in Design*. A.C.T., Australia: RAIA Education Division.

Templeton, S. 1990. Some observations of black-headed python behavior in captivity. *Thylacinus*, **15**, 11-12.

Underwood, G. & A.F. Stimson. 1990. A classification of pythons (Serpentes, Pythoninae). *The Zoological Society of London*, **1**, 565-603.

Van Mierop, L.H.S., T. Walsh & D.L. Marcellini. 1982. Reproduction of *Chondropython viridis* (Reptilia, Serpentes, Boidae). In: *6th Annual Reptile Symposium on Captive Propagation and Husbandry* (Ed. by R. Hahn), pp. 256-274. Thurmont, Maryland: Zoological Consortium Inc.

Wagler, J. 1830. *Naturliches System der Amphibien, mit vorangehender Classification der Saugthiere und Vogel*. J.G. Cotta, Munchen, Stuttgart & Tubingen.

Walsh, T. 1979. Further notes on the husbandry, breeding, and behavior of *Chondropython viridis*. In: *3rd Annual Reptile Symposium on Captive Reproduction and Husbandry* (Ed. by R. Hahn), pp. 102-111. Thurmont, Maryland: Zoological Consortium Inc.

Weigel, J. 1990. *The Snakes of Southeast Australia*. Gosford, New South Wales, Australia: The Australian Reptile Park.

Weigel, J. & T. Russell. 1993. A record of a third specimen of the Rough-Scaled Python, *Morelia carinata*. *Herpetofauna*, **23**, 1-5.

Wells, R.W. & C.R. Wellington. 1984. A synopsis of the Class Reptilia in Australia. *Australian Journal of Herpetology*, **1**, 73-129.

Wells, R.W. & C.R. Wellington. 1985. A classification of the Amphibia and Reptilia of Australia. *Australian Journal of Herpetology*, **1**, 1-61.

Werner, F. 1909. Uber neue oder seltene Reptilien des Musee Royal d'Histoiore Naturelle de Belgique in Brussell. *Geographie und Biologie der Thiere*, **28**, 263-288.

Werner, F. 1936. Nova species boidarum. *Annales Musei Nationalis Hungarici, Zoologica*, **30**, 105.

Whitlock, F. 1923. A trip to the Fortescue River and Hamersley Ranges, North-west Australia. *Emu*, **22**, 259-273.

Wiley, E.O. 1981. *Phylogenetics*. New York: John Wiley & Sons.

Williams, D. 1992a. Natural incubation by a children's python, *Liasis maculosus*, (Serpentes: Boidae), under public display conditions. *Sydney Basin Naturalist*, **1**, 79-80.

Williams, D. 1992b. Notes on the carpet python, *Morelia spilota variegata* (Gray, 1842), from the Mackay Region of coastal Queensland. *Sydney Basin Naturalist*, **1**, 79-80.

Wilson, S.K. & D.G. Knowles. 1988. *Australia's Reptiles: A Photographic Reference to the Terrestrial Reptiles of Australia*. Sydney, Australia: William Collins Pty Ltd.

Worrell, E. 1963. *Reptiles of Australia*. Angus and Robertson Ltd., Sydney.

Zacharias, H.C.E. 1897. Die phylogenese der kopfschilder bei den boiden. *Zoologischen Jahrbucher, Abtheilung fur Systematik, Geographie and Biologie der Thiere*, **10**, 56-90.

Index

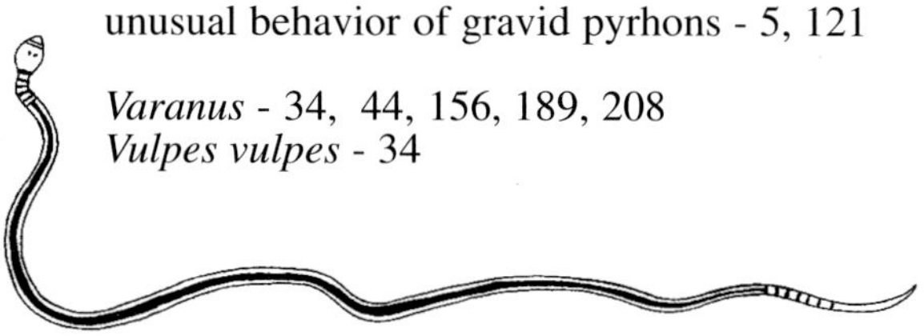